TO:
THE AMERICAN UNIVERSITY

FROM:
THE CLASS OF 1932 LIBRARY
ENDOWMENT FUND

International
Agricultural Trade

Westview Replica Editions

The concept of Westview Replica Editions is a response to the continuing crisis in academic and informational publishing. Library budgets for books have been severely curtailed. Ever larger portions of general library budgets are being diverted from the purchase of books and used for data banks, computers, micromedia, and other methods of information retrieval. Interlibrary loan structures further reduce the edition sizes required to satisfy the needs of the scholarly community. Economic pressures on the university presses and the few private scholarly publishing companies have severely limited the capacity of the industry to properly serve the academic and research communities. As a result, many manuscripts dealing with important subjects, often representing the highest level of scholarship, are no longer economically viable publishing projects--or, if accepted for publication, are typically subject to lead times ranging from one to three years.

Westview Replica Editions are our practical solution to the problem. We accept a manuscript in camera-ready form, typed according to our specifications, and move it immediately into the production process. As always, the selection criteria include the importance of the subject, the work's contribution to scholarship, and its insight, originality of thought, and excellence of exposition. The responsibility for editing and proofreading lies with the author or sponsoring institution. We prepare chapter headings and display pages, file for copyright, and obtain Library of Congress Cataloging in Publication Data. A detailed manual contains simple instructions for preparing the final typescript, and our editorial staff is always available to answer questions.

The end result is a book printed on acid-free paper and bound in sturdy library-quality soft covers. We manufacture these books ourselves using equipment that does not require a lengthy make-ready process and that allows us to publish first editions of 300 to 600 copies and to reprint even smaller quantities as needed. Thus, we can produce Replica Editions quickly and can keep even very specialized books in print as long as there is a demand for them.

About the Book and Editors

International Agricultural Trade:
Advanced Readings in Price Formation,
Market Structure, and Price Instability
edited by Gary G. Storey, Andrew Schmitz, and Alexander H. Sarris

This book examines the institutions and market structures of
international agricultural trade, the formation of commodity prices,
and the effects of price instability on international markets.
Emphasizing the need for a proper understanding of the role of
speculative storage and futures markets, the authors discuss the
factors that can affect price movements, including weather, exchange
rate movements, and changes in monetary policies and interest rates.
Their subsequent study demonstrates how certain institutions facilitate international trade while benefiting producers of primary commodities, how commodity options facilitate the exchange of goods,
and how major marketing boards can benefit from participation in
futures trading. The authors also show how governments can reduce
treasury spending without affecting producer welfare by combining
deficiency payments with farmer-held reserves rather than relying on
price supports. Finally, they consider the role of cartels and
other forms of market power, maintaining that traditional stabilization policies may be inappropriate in the context of imperfect
markets.

Gary G. Storey is professor of agricultural economics at the
University of Saskatchewan. Andrew Schmitz is professor in the
Department of Agricultural and Resource Economics at the University
of California at Berkeley. Alexander H. Sarris is professor of
economics at the University of Athens, Greece.

International Agricultural Trade

Advanced Readings in Price Formation, Market Structure, and Price Instability

edited by Gary G. Storey,
Andrew Schmitz, and Alexander H. Sarris

Westview Press / Boulder and London

Copyright © 1984 by Westview Press, Inc.

Published in 1984 in the United States of America by
    Westview Press, Inc.
    5500 Central Avenue
    Boulder, Colorado 80301
    Frederick A. Praeger, Publisher

Library of Congress Cataloging in Publication Data
Main entry under title:
International agricultural trade.
    (A Westview replica edition)
    1. Produce trade--Addresses, essays, lectures. I. Storey, Gary G.
II. Schmitz, Andrew. III. Sarris, Alexander H.
HD9000.5.I564   1984   382'.41   83-10336
ISBN 0-86531-955-3

Printed and bound in the United States of America
10 9 8 7 6 5 4 3 2 1

Contents

Tables

Figures

Preface

Agricultural trade has become an integral part of
world agriculture. During the 1970s, the real growth
in world agricultural trade was phenomenal. For
example, the value of U. S. agricultural exports alone
increased more than fivefold during this period.

Trade in agricultural goods has many peculiarities,
most of which fit in the tariff and nontariff barrier
categories. Given the existence of many distortions,
including market imperfections, the ideal world of
free or liberalized agricultural trade is nonexistent.
This book focuses largely on the question of how
world agricultural trade functions when it is far
removed from the theoretical free-trade model.

In April, 1978, a small group of West Coast
agricultural economists (Hillman, Josling, Sarris,
Schmitz, King, and McCalla) met to form what is now
called the International Trade Consortium which is
financed, in part, by the U. S. Department of
Agriculture and Agriculture Canada. One of the
products of this project was a book published in 1979
by A. F. McCalla and T. E. Josling (editors),
Imperfect Markets in Agricultural Trade, Allenheld,
Osmun and Co., 1981. In the same vein, our current
book is a result of an International Trade Consortium
meeting held in Berkeley, California, in the early
1980s.

This book focuses heavily on the policy process
operating in a world of distorted and noncompetitive
markets. Many of the chapters combine a heavy dose
of theory and quantitative methods; hence, some of
the chapters may prove to be difficult reading for
the nonspecialist. However, the advanced chapters
also contain summaries which, hopefully, will prove
to be useful.

Many organizations and individuals contributed
to this project. However, we would like to thank
especially the Economic Research Service, U. S.
Department of Agriculture; Agriculture Canada; and

Mrs. Eileen Bonnor and Mrs. Gary Storey, for their typing and editorial assistance.

Gary G. Storey
University of Saskatchewan

Andrew Schmitz
University of California at Berkeley

Alexander H. Sarris
University of Athens
Greece

Part I

Price Formation in Agricultural Commodity Markets

1
Real and Monetary Determinants of Non-Oil Primary Commodity Price Movements

Enzo R. Grilli and Maw-Cheng Yang

INTRODUCTION

Global and sectoral primary commodity price movements since World War II have been subjected to relatively few systematic analyses. Those available usually concentrate on a sub-period (especially the past ten years), and are geared either towards the explanation of a particular commodity cycle or cycles (e.g., Radetzki (1974) and Cooper and Lawrence (1975)), or towards commodity price forecasting (e.g., Ray and Timm (1980) and OECD (1981)). We are only aware of three quantitative analyses of global and sectoral commodity price movements over the past three decades: one by IMF, one by World Bank and one by Bank of England staff (e.g., Goreux (1975); Grilli (1975) and Enoch and Panic (1981)).

There is no want, on the other hand, of analyses of behavior of single commodity prices. Labys (1978) has recently made a comprehensive, if by no means exhaustive, survey of this literature. Some very interesting attempts have, moreover, also been made to analyze and project short-term price movements of primary commodities using reduced forms of dynamic disequilibrium models of price adjustments (e.g., Hwa (1979)) and ARIMA models (e.g., Chu (1978)). Mixed stock/flow models of single commodity markets, specified in continuous time, have also become available (e.g., Wymer (1973) and Richard (1978)).

The paucity of quantitative analysis of global and sectoral commodity price movements over the long term probably results from the difficulties encountered in

Enzo R. Grilli, Segretario Generale P.E. Ministero del Bilancio e Della Programmazione Economica, Italy and Maw-Cheng Yang, Economist, World Bank, Washington, D. C.

estimating price formation models, given that inventory
data are not available and adequate proxies are
difficult, if not impossible, to find for the range of
commodity aggregates that need to be covered. The gap
between the specification and estimation of price
models, in a multi-commodity dimension, is quite large:
dynamic disequilibrium models cannot be empirically
tested and even equilibrium models can only "be made"
dynamic by using fairly mechanical procedures. For
these (and perhaps other) reasons, analysts have
preferred to take a single commodity approach.
Although rich in detail and sometimes elegant in
construction, this has not produced answers to some
very important general questions.

One of these questions relates to the overall
influence of the business cycle in industrial countries
on the relative prices of primary commodities (primary
commodity/manufactured product prices), which for a
certain group of developing countries--the non-oil and
non-NICs[1]--comes close to their barter terms of trade.
A second question is related to the impact of the
growth of international liquidity in the 1970s on
relative commodity prices and its effect on the terms
of trade of commodity exporting and commodity importing
countries. A third less general, but still quite
important, question relates to the impact of short-term
interest rate changes on commodity prices. Finally,
the question of how exchange rate instability affects
commodity prices is also of interest.

In this paper we analyze global and sectoral
primary commodity price behavior over the past three
decades using simple equilibrium models of price
determination which are still helpful in providing some
answers to these questions. The empirical tests of
these models yield results which are, in our view,
sufficiently robust to justify the effort made. The
trade-offs between simple equilibrium market models
for the analysis of aggregate commodity price behavior
to provide some answers to the questions posed, and
more sophisticated disequilibrium models which have in
practice a limited applicability and have to be tested
in a single commodity dimension, are not yet clear.
Hopefully, as research in this field proceeds, both the
size and the relevance of the trade-offs will become
more apparent. At this stage we simply intend to bring
into the open the results obtained using our approach,
possibly starting a debate on these issues which,
despite their importance, have so far received little
attention in the economic literature.

COMMODITY PRICE MOVEMENTS: THE POST WORLD WAR II
EXPERIENCE

Overall, non-oil primary commodity prices[2] more

than trebled in nominal terms between 1948 and 1980.
Food prices set the norm, surpassed by metal prices,
while agricultural non-food price growth lagged behind.
Relative to those of manufactured products[3], non-oil
commodity prices declined by about seven percent during
this period. Food prices deviated little from this
average; metal prices showed a modest increase during
the period, while there was a fifteen percent decline
in agricultural non-food prices. These comparisons are
illustrated in Figures 1, 2, 3 and 4.

The choice of any specific time period for
measuring price movements is always difficult. We have
chosen 1948 as the beginning year to avoid the problems
connected with the commodity price boom caused by the
Korean War. No drastic change in the long term trends
of non-oil primary commodity prices, relative to those
of manufactures, is apparent between 1948 and 1980.
This applies to the overall price index, as well as to
the prices of food and metal products. The only
exception is with regard to agricultural non-food
prices. Here a falling trend emerging in the 1950s and
continuing in the 1960s appears to have levelled off in
the 1970s, as a consequence of the cost and price
increases of the petroleum-based substitutes of these
products (synthetic fibers, polyolefins and synthetic
rubbers).[4]

Looking at the movements of non-oil commodity
prices relative to those of manufactures during this
period, two peaks and two troughs are apparent in the
aggregate prices: prices reached relative maxima in
1950 and 1974, and fell to relative minima in 1962 and
1978. The first peak is well known and connected to
the effects of the Korean War. The second is also
commonly, if somewhat mistakenly, identified with the
so called "oil crisis", which in itself seems to have
had only a limited effect on commodity prices. Only
food prices, in fact, peaked strongly in 1974; non-
food prices had peaked in the previous year and were
already falling in relative terms in 1974. Metal
prices, on an annual basis show a peak in 1974, but
actually their increase did not last much beyond the
first quarter of the year.

The troughs of 1962 and 1978 are slightly more
difficult to explain. In contrast to 1958 and 1975,
when demand factors clearly played a major and uniform
role in the downturn of commodity prices--as witnessed
by the syncronic downturn in the industrial output
of OECD countries--the influence of demand on both the
1962 and 1978 commodity price troughs is less clear.
Neither year was characterized by boom demand
conditions. In 1962 the growth in industrial output[5]
was decelerating for the second consecutive year in
Japan and was barely recovering in Western Europe,
after the drastic fall of the previous year. Only in

6

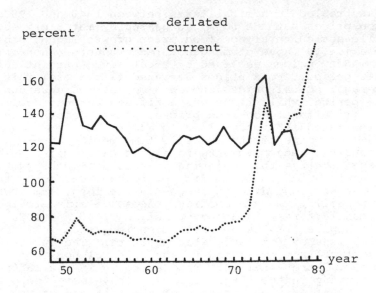

Figure 1   Commodity Price Index - 30 Commodities

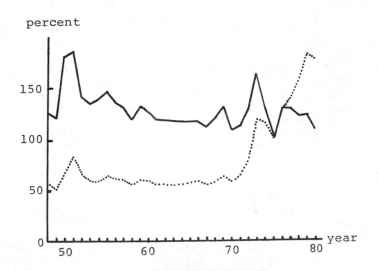

Figure 2   Commodity Price Index - Non-Food

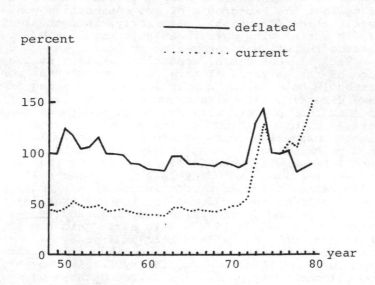

Figure 3   Commodity Price Index - Food

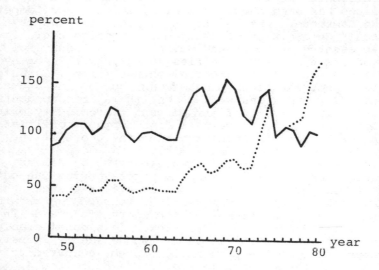

Figure 4   Commodity Price Index - Metals

the U. S. was industrial output growing fast in 1962,
if against the background of two consecutive years of
deceleration (Figure 5). Similarly, industrial output
growth was slowing down in 1978 in both the U. S. and
Western Europe, but recovering in Japan.

Yet, if weak demand conditions explain at least
some of the softness of metal and agricultural non-food
prices in both years, supply factors operating in the
same direction were also at work in both 1962 and 1978.
In 1962 food prices, relative to manufactured goods,
were at a local minimum at the end of a period of
eight consecutive years of steady decline. In 1978
food prices also dipped strongly, but the decline was
much less general than in 1962: only tropical beverage
and sugar prices fell drastically from their 1977 peak,
while cereal and meat prices increased appreciably.
The effects of the 1975 coffee frost in Brazil--which
had also pushed up the prices of cocoa and tea--were
coming to an end in 1978, while sugar prices were at
their cyclical low, after their 1974 peak.

During the entire post-war period, and particularly
from the mid-1950s onwards, the cyclical movements of
non-oil commodity prices--relative to those of
manufactures--are explained to a considerable extent by
those of industrial activity in OECD countries
(Figures 6, 7, 8 and 9). The importance of demand
factors in the determination of short-term commodity
prices has been shown by Grilli, Goreaux and Cooper
and Lawrence. Whether the best proxy for the key
"real" determinant of the commodity price cycle is
represented by the trend deviations of industrial
production in OECD countries (e. g., Grilli) or by the
ratio of actual to potential output (e. g., Goreaux)[6],
its impact is shown to be strong, but on the aggregate
commodity index, as well as on its major components,
explaining forty to fifty percent of commodity price
variations.[7] These results find further confirmation
in more recent analyses covering the whole of the
1970s (e. g., Enoch and Panic).

The focus on the influence exerted by demand
factors on relative commodity prices during the entire
post-war period obscures some important changes that
have taken place in the past ten years. The growth of
industrial production--and therefore of demand--in OECD
countries during the 1970s slowed down considerably
in comparison to the previous two decades: industrial
output grew at 2.9 percent per annum, compared to the
7.1 percent per annum in the 1960s and to the 4.3
percent per annum in the 1950s. Despite the relatively
faster growth of industrial production that occurred in
the 1970s in centrally planned economies and developing
countries, world aggregate industrial output increased
at only slightly higher rates than that of the OECD
countries: at about 3.5 percent per annum in the 1970s,

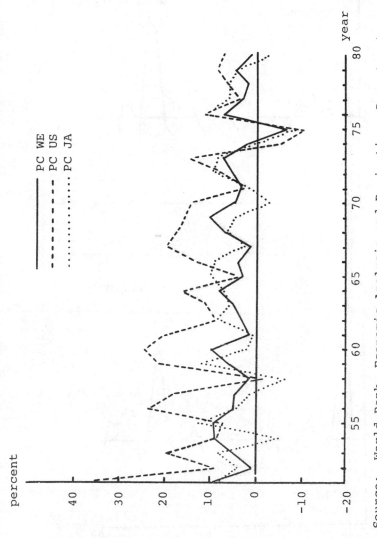

Source: World Bank, Economic Analysis and Projections Department, Commodities and Export Projections Division

Figure 5   Industrial Production Index (% Change from Previous Year)

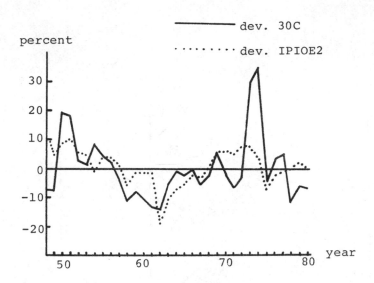

Figure 6 OECD Industrial Production and
Commodity Price (% deviation from trend)

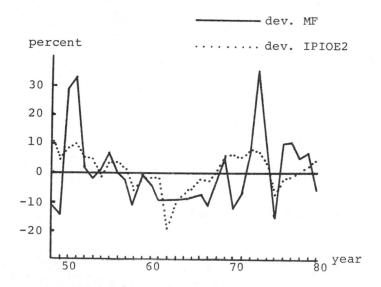

Figure 7 OECD Industrial Production and Non-
Food Price (% deviation from trend)

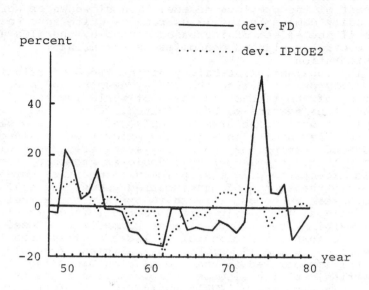

Figure 8   OECD Industrial Production and Food
Price (% deviation from trend)

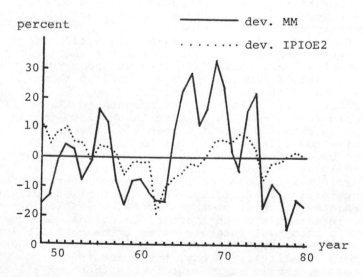

Figure 9   OECD Industrial Production and Metal
Price (% deviation from trend)

against the 6.7 percent of the 1960s and the 4.5
percent of the previous decade. The slowdown of world
commodity demand in the 1970s remains quite apparent,
even if the effect of increased demand by the NICs and
the centrally planned economies is taken into
consideration.

The substantial stability of the trend of relative
commodity prices during the 1970s, despite the near
halving of the rate of increase of world industrial
demand thus remains to be explained. Given that
suppliers of manufactures probably have, on the whole,
greater flexibility to adjusting production to slowdowns
in demand than primary commodity suppliers, one could
expect that, ceteris paribus, a slower growth of demand
would affect the prices of primary commodities more
severely than those of manufactured goods. Several
hypotheses can be put forward to explain the actual
outcome: a) the likely shifts to the left of the long-
term marginal cost curve of most commodities caused by
the more than quadrupling of real crude oil prices that
took place in the 1970s; b) the policies of supply
control implemented in some developing producing
countries; c) the slowdown of production capacity
growth in some key minerals caused by sharply reduced
expectations concerning future demand, by the greater
political risks of foreign mining investments in
developing countries, and by the technological
developments which have caused mining and mineral
processing projects to become lumpier, and therefore
inherently riskier over time; and d) the increase in
the short and long-term costs of production of
petroleum-based synthetic substitutes to agricultural
commodities. All these factors have undoubtedly played
a role, even if the empirical evidence on those
summarized in a) and c) is scanty.

However, if we look at the changes in the world
economic environment which have taken place in the
1970s, we cannot ignore the possible effects on
relative commodity prices that may have come from:
a) the breakdown of the fixed exchange rate system in
the early 1970s and the subsequent greater volatility
of foreign exchange markets; b) the staggering growth
of international liquidity, both official and private;
and related to that, c) the sharp acceleration of
world inflation and the increase in nominal and more
recently also real interest rates. The possible impact
of these "monetary" factors on commodity prices has so
far received little, if any, attention in the
literature. While it is conceivable that not all the
abovementioned phenomena may have had marked effects
on relative commodity prices, their influence on both
trend and cycles needs to be studied if the changes
in commodity markets in the 1970s, and their influence
on the world economy, are to be understood more fully.

Industrial production also fluctuated more in the 1970s than in the previous two decades. This too might have affected the primary commodity price cycle, given the differences that appear to exist in the price formation mechanisms between commodities and manufactured products. One could, in fact, expect that for the same variation in demand, commodity prices fluctuate more than manufactured good prices, because of the more limited ability of commodity suppliers to adjust production to demand changes, with respect to that of suppliers of manufactured goods. Both industrial production in OECD countries and relative commodity prices fluctuated more violently in the 1970s than in the 1960s and the 1950s (Table 1).[8]/

TABLE 1
Coefficients of Variation of OECD Industrial Production and Commodity Prices[a]

	1951–1960	1961–1970	1971–1980
OECD industrial production[b]	0.67	0.47	1.40
Commodity price index (total)	0.096	0.056	0.156
Food price index	0.108	0.054	0.195
Non-food Ag. price index	0.131	0.049	0.137
Metal price index	0.097	0.177	0.152

Source: Computed

[a]Relative to those of manufactured products.
[b]Corrected for exponential trend growth.

It would nevertheless be simplistic to attribute to "real" factors all the increment in commodity price fluctuations. With the rise of interest rates, and the opportunity costs of carrying stocks, industrial processors of primary products might have tended to reduce the level of working inventories relative to expected demand, generating stronger than usual inventory accumulation or decumulation over the demand cycle. Moreover, exchange rate instability might have created temporary shifts between liquid and real assets, including commodities—and vice versa—during the 1970s, increasing the short-term volatility of commodity markets.

The growth of international liquidity may have also influenced commodity price movements in the 1970s. The effect could be twofold. To the extent that commodity producing countries have greater and easier access to international liquidity (which in turn should be a positive function of its growth), "distress sales" of primary commodities at times of low market demand and prices can be postponed or avoided and export supply of commodities can be better controlled by government authorities. The growth of "official" liquidity could, therefore, have had a positive influence on commodity prices reducing their instability and possibly influencing their trends. At the same time, the growth of privately held international liquidity poses its holders with a placement choice. In times of high inflation this choice must be strongly influenced by the goal of maintaining at least the "real" value (or purchasing power) of the holdings. Purchases of commodities as a store of value for liquid assets internationally held cannot but add to the industrial demand for these products and have a positive influence on their prices. Across commodity groups, it is to be expected that, if this liquidity effect exists, it should be relatively stronger on the prices of those storable commodities that have shown a tendency to improve over time in real terms (e. g., metals).

DETERMINANTS OF ANNUAL COMMODITY PRICE MOVEMENTS:
MODEL SPECIFICATION AND EMPIRICAL ESTIMATES

The framework within which aggregate commodity price relationships are here formulated assumes that commodity markets are competitive. This is a fairly common and probably realistic assumption, given that we are considering aggregate commodity price relationships. We further assume that markets clear and that inventories, where they exist, do not vary significantly in the aggregate from year to year. At the level of a single commodity this assumption would be appropriate only for highly perishable products (meat, bananas, citrus fruits, etc.). Inventories do exist in most commodity markets. Their levels and changes do play a role in the market price adjustment process.

However, given the scarcity and non-uniformity of inventory data and the aggregation problems that have to be faced in dealing with clusters of commodities for each of which information on world production and consumption is available only in quantities, stock-adjustment disequilibrium models, cannot be easily used to analyze aggregate commodity price relationships. They can be specified, but not fully tested in a multi-commodity dimension. When proxies, at least for consumer stocks, can be constructed (e. g., OECD), their usefulness is limited to the analysis of particular commodity aggregates--for example, industrial raw materials, where imports are very important relative to total consumption.

Moreover the use of a competitive market equilibrium model for the analysis of aggregate commodity price relationships using annual data may not be unduly restrictive, if one considers that for clusters of commodities inventory movements should be, at least in part, mutually offsetting, and that annual aggregate commodity price indexes should represent near equilibrium values. In the end, whether this is a reasonable assumption, is as much an empirical as a theoretical question.

The single equation (annual) model used here to test the relative importance and significance of various commodity price determinants can be thought of as the reduced form of a competitive market equilibrium model where demand for a commodity or group of commodities is:

$$Qd = f(CPI / AMPIG, IPIOE, ILR) \qquad (1)$$

where: CPI   = index of own prices (in U. S. dollars)
AMPIG = index of other prices (in U. S. dollars)
IPIOE = index of OECD industrial production
ILR   = world liquidity indicator

and the supply is:

$$Qs = f(CPI / AMPIG, T) \qquad (2)$$

where: T = Time trend

and the equilibrium condition is:

$$Qd = Qs \qquad (3)$$

If (1) and (2) are specified in linear form, the reduced form equation for price which can be derived is:

$$\frac{CPI}{AMPIG} = \alpha_0 + \alpha_1 \ IPIOE + \alpha_2 \ ILR + \alpha_3 \ T \qquad (4)$$

in which α_0, α_1, α_2 and α_3 are summary expressions of the parameters of the reduced form in terms of those of equations (1) and (2). Equation (4) is the one used in the econometric estimation of the relative price relationships for all commodities (CPI3OD), food commodities (CPIFDND), non-food commodities (CPINFND) and metals (CPIMMND).

In equation (1) the sign of the coefficient of the industrial production variable (IPIOE) is expected to be positive; those of the own price and other price variables (CPI and AMPIG) are expected to be negative and positive respectively.[9]/ The sign of the coefficient of the world liquidity indicator variable (ILR) is also expected to be positive. In equation (2) the signs of the coefficients of CPI and AMPIG are expected to be positive and negative respectively, while that of the time trend variable (T) is expected to be positive. This variable is a proxy for technological and other changes in the production of commodities that positively influence their supply.

Two international liquidity variables were constructed (IL1 and IL2). The first one embodies the traditional concept of international liquidity, as the U. S. dollar value of foreign exchange reserves and SDRs held by national monetary authorities, complemented by the value of their reserve position in the IMF.[10] IL1 is in essence a measure of international reserve holdings by national monetary authorities. The second one, on the contrary, embodies the less orthodox concept of international monetary base (IMB), akin to that of a country's national monetary base. IL2 consists of the U. S. dollar value of gold, net of private uses and gold deposit at the IMF, the reserve position of each member country in the IMF plus credits granted by the IMF, SDRs, credit lines in dollars granted by the U. S. Federal Reserve to non-residents, credit lines in currencies other than dollars granted by other central banks to non-residents, dollars created (destroyed) by deficits (surplus) in the U. S. balance of payments (measured on IMB basis) and other convertible currencies which function as international monetary base (e. g., Fratianni and Savona (1972) and 1973)).

Both indicators of world liquidity were deflated by the value of world trade. The distribution of IL2 between official and non-official holders was also used to create an indicator of international liquidity giving the share of the total held by non-monetary authorities (ILR). The subsidiary hypothesis is that the greater the share of international liquidity held in private hands, the greater the possibility of liquidity spillovers in commodity markets for purchases of commodities to be kept as stores of value.[11]

The results of the econometric estimates of the annual price model applied to the aggregate commodity index and to the three sub-indexes are shown in Table 2. The estimation periods are 1948-80 and 1957-80. The shorter period estimates were made necessary by the availability of IMB statistics only from 1957.[12]

Industrial production is shown to have had a consistently positive and highly significant impact on relative commodity prices. The influence of the IPIOE variable--an indicator of business cycle condition--is also, as expected, stronger on metal and non-food commodity prices than on those of food products.

TABLE 2
Relative Commodity Price Equations (Annual Data)[a]

	Independent Variables				
	Constant	IPIOE	ILR	T	DYM74
30 commodities price index (CPI30D)	98.69 (16.16)	1.92* (5.25)		-7.13* (-5.50)	19.23* (2.95)
	28.19 (1.87)	2.29* (5.36)	0.01 (0.02)	-8.93* (-5.16)	17.46* (3.09)
	28.28 (2.00)	2.29* (5.58)		-8.93* (5.30)	17.47* (3.18)
Food price index (CPIFDND)	91.04 (14.93)	1.49* (3.86)		-5.56* (-4.12)	31.74* (4.05)
	42.42 (2.20)	1.65* (2.95)	-0.39 (-1.00)	-5.80* (-2.58)	29.68* (3.93)
	34.81 (1.84)	1.64* (3.00)		-6.07* (-2.70)	28.11* (3.88)

TABLE 2 (Cont'd)

	Test Statistics				Period
	p	Adj R^2	D-W	S.E.E	
30 commodities price index (CPI30D)	0.48	0.59	2.09	6.98	1948-80
	0.52	0.69	1.83	6.08	1957-80
	0.52	0.71	1.83	5.92	1957-80
Food price index (CPIFDND)	0.37	0.55	2.02	8.08	1948-80
	0.46	0.59	1.89	7.92	1957-80
	0.54	0.59	1.83	7.88	1957-80

TABLE 2 (Cont'd)

		Independent Variables			
	Constant	IPIOE	ILR	T	DYM75
Non-food price index (CPINFND)	122.47 (11.19)	3.12* (4.77)		-11.93* (-5.15)	-12.60 (-1.09)
	7.17 (0.30)	3.70* (6.58)	0.14 (0.31)	-15.12* (-6.27)	-15.75* (-2.22)
	138.02 (9.29)	3.19* (5.49)		-12.86* (-5.33)	
Metal price index (CPIMMND)	128.09 (4.73)	2.86* (4.48)		-11.36* (-4.29)	17.53* (2.10)
	9.08 (0.37)	2.84* (4.22)	1.22* (2.41)	-12.67* (-4.59)	15.90 (1.82)
	37.62 (1.31)	2.96* (4.18)		-12.62* (-4.12)	17.10 (1.89)

TABLE 2 (Concluded)

	Test Statistics				Period
	p	Adj R^2	D-W	S.E.E	
Non-food price index (CPINFND)	0.48	0.45	1.89	12.41	1948-80
	0.71	0.65	1.55	8.33	1957-80
	0.54	0.56	1.81	8.77	1957-80
Metal price index (CPIMMND)	0.82	0.48	1.70	10.42	1948-80
	0.58	0.62	1.70	9.66	1957-80
	0.70	0.53	1.60	10.63	1957-80

Source: Computed

[a]Estimated by OLS, using the Cochrane-Orcut correction for first order serial correlation.
(t values in parentheses).
*Significant at the 95% confidence level or above.

The time trend variable shows a negative sign and
is consistently significant. The negative sign of this
variable implies that the effect of technological and
other changes affecting productivity is stronger on the
prices of primary commodities than on those of
manufactures. Given the less competitive nature of
manufactured product markets with respect to those of
primary commodities, this result is not surprising.
Across primary commodity groups the negative value of
the time trend is shown to be larger in the case of
metal and non-food commodity prices than in that of
food prices. These results conform to the notion that
productivity changes over the past thirty years have
been more rapid in metals and non-food agricultural
raw materials than in food production.

Of the international liquidity indicators only ILR
(the share of international liquidity held in private
hands) turned out to have a positive sign and to be
significant in explaining relative commodity price
movements, if only in the case of metals. Neither IL1
nor IL2, however deflated, yielded any satisfactory
results. The lack of significance of IL1 (an indicator
of the value of official reserves) could mean that on
the aggregate there is no firm relationship between the
growth of international liquidity traditionally defined
and relative commodity price changes: either "distress"
sales by government authorities are not in the
aggregate a sizeable factor in commodity markets or the
growth of international reserves per se does not
prevent them.

The significance of ILR--if, as expected, only in
the case of metal prices--seems on the contrary to
indicate that the larger the share of the international
monetary base held by non-official authorities, the
greater the spillover of this liquidity to commodity
markets, particularly to metal markets. Recalling
that metal prices have in the aggregate increased over
time relative to those of manufactured products,
diversification of liquid assets into some metal
assets (particularly nickel and tin) is indeed a good
way of hedging against inflation in the medium term.13/

The existence of a positive link between the
growth of international liquidity and relative
commodity prices (until now postulated by only a few
analysts and usually sketched out in terms of
speculative behavior (e. g., Biasco (1979)) finds
empirical support in our analysis. It should be
stressed, however, that what appears to influence
commodity markets is not the growth of official
reserves, but that of the share of international
monetary base held in private hands. This influence,
moreover, seems to be limited to the markets for metals.

Finally, the dummy variable for the year 1974

(DMY74), added to the price equations to test the importance of "special factors" on the developments of relative commodity prices in that year, shows consistently the right sign significant only for the food and aggregate price indexes. This finding runs against the widespread notion that expectations generated by the oil crisis of late 1973, pushed up commodity prices in 1974, despite weak industrial demand.

In fact the relative prices of non-food agricultural raw materials and metals do not appear to have been significantly influenced by the turmoil generated by OPEC action during 1974.[14] On the contrary, relative food prices, and because of their weight in it, the aggregate price index, were positively affected in 1974 by special circumstances. These, however, can be traced not to the oil crisis of late 1973, but to the shortfall in world production of cereals that took place in that year, at a time when world stocks were low because of the extremely poor crop of 1972.[15] Traditional, random supply factors, related to poor weather conditions, were mostly responsible for the food price boom of 1974.[16]

A test on asymmetry in the response of relative commodity prices to changes in demand conditions was conducted. The deviation from trend of each commodity price index was regressed on deviation from trend of the index of industrial production in OECD countries (IPIOE). A dummy variable of negative deviation (IPIOE*DN) was created and entered in the regression equations. As shown by Goldstein (1977), if the coefficient of IPIOE*DN is significantly different from zero, then it indicates that an asymmetric response exists. The results of these tests are shown in Table 3. They indicate that a clear assymetric response is only present in the case of non-food prices. Positive deviation from trend of the index of industrial production in OECD countries appears to influence non-food prices in the upward direction more than negative deviations from trend do in the opposite direction.

The non-asymmetric response of metal prices would seem to indicate that on the whole the metal industries have operated during this period below their productive capacity and, therefore, that changes in the demand curve for these products have put the short-term supply curve well below its point of (upward) inflection.

Two tests were performed on the stability of the estimated coefficients of the first and the second price equations shown in Table 2 under the commodity headings. The first test involved the calculation of the cumulative sum of residuals (CUSUM) and the

TABLE 3
Asymmetry Tests on the Economic Activity Variable (IPIOE)

Independent Variables	Dependent Variables			
	DEV.30C	DEV.FD	DEV.NF	DEV.MM
1948-1980				
DEV.IPIOE	2.79* (4.08)	2.45* (3.20)	4.72* (4.16)	2.47* (2.35)
DEV.IPIOE*DN[a]	-1.84 (-1.70)	-1.83 (-1.52)	-3.71* (-2.07)	0.16 (0.09)
1948-1965				
DEV.IPIOE	1.40 (1.29)	0.80 (1.04)	13.72* (3.41)	-2.21 (-1.27)
DEV.IPIOE*DN[a]	0.32 (0.21)	1.06 (0.90)	-12.38* (-2.68)	4.92 (1.99)
1965-1980				
DEV.IPIOE	3.10* (3.58)	2.88* (2.88)	4.17* (4.39)	2.75* (2.54)
DEV.IPIOE*DN[a]	-2.36 (-1.23)	-4.27 (-1.90)	0.11 (0.05)	0.23 (0.10)

TABLE 3 (Cont'd)

Independent Variables	Dependent Variables			
	DEV.30C	DEV.FD	DEV.NF	DEV.MM
1957-1980				
DEV.IPIOE	2.94* (4.46)	2.54* (3.07)	4.29* (4.61)	2.84* (2.71)
DEV.IPIOE*DN[a]	-1.96 (-1.76)	-1.85 (-1.31)	-3.49* (-2.15)	-0.88 (-0.49)

Source: Computed

[a]DN is a dummy variable, taking the value of one when the deviation from trend in IPIOE is negative and the value zero when deviation from trend in IPIOE is positive.

(t values in parentheses)

*Significant at the 95% confidence level or above.

cumulative sum of squares residuals (CUSUMSQ) by
recursive regression. The second test involved the
recomputation of the coefficients on successive
intervals of ten - eleven years by moving regression
(Brown, Durbin and Evans, 1975). The numerical results
of the CUSUM and CUSUMSQ tests are summarized in
Table 4, which also contains the significance values of
the tests at the one percent and ten percent levels for
the two sets of price equations for which the analysis
of coefficients' stability is performed. The
"homogeneity statistics" derived from the moving
regressions for the same sets of price equations are
given in Table 5.

The numerical CUSUM tests performed on both
forward and backward recursive regressions tend to
indicate that the coefficients of all the price
equations remained constant during the 1948-1980 period.
The null hypothesis--that the regression coefficients
have remained constant over time--cannot be rejected
for any of the price equations at the ten percent
level of significance. The test results hold for both
the forward and backward recursions. Examination of
the plots also shows randomness in the recursive
residuals, confirming the numerical test results.
According to the CUSUMSQ tests, however, the null
hypothesis can be uniformly rejected at the ten percent
confidence level for all price equations except metals.
At the one percent confidence level the null hypothesis
can be rejected uniformly for the food and non-food
price equations.

The same tests performed on the second set of price
equations estimated over the 1957-1980 period and
containing the ILR variable in addition to IPIOE, DYM74
and the time trend, lead to similar results. The
numerical CUSUM tests are uniform across backward and
forward recursions in showing that the null hypothesis
cannot be rejected for any of the price equations.
According to the CUSUMSQ tests, the null hypothesis
can be rejected at both the ten percent and one
percent confidence level for the aggregate and the food
price equations, but not for the non-food and the metal
price equations.[17/]

Given the nature of the two tests,[18/] the CUSUM test
results are probably more relevant, since they are
specifically designed to ascertain the existence of a
systematic change in the structure of the price
equations. From this standpoint, the CUSUM tests would
tend to indicate that there has not been any major
structural change over the years in the relative
commodity price formation mechanism.

The additional constancy tests obtained from the
moving regressions support this conclusion. The
"homogeneity statistics" shown in Table 5 consistently
show that the null hypothesis cannot be rejected at the

TABLE 4
Constancy Tests on the Estimated Coefficients of Price
Equations Using Recursive Regression

	CUSUM Test[a] Values	CUSUMSQ Test[b] Values
1948-1980 (32 observations, 4 parameters)[c]		
CPI30D: BRR[d]	0.42764	0.26427
FRR	0.75336	0.30096
CPIFDND: BRR	0.43772	0.31147
FRR	0.63388	0.34233
CPINFND: BRR	0.42622	0.49368
FRR	0.75494	0.33886
CPIMMND: BRR	0.69412	0.20155
FRR	0.35250	0.16288
1957-1980 (23 observations, 5 parameters)[c]		
CPI30D: BRR	0.49202	0.63148
FRR	0.32567	0.57988
CPIFDND: BRR	0.53097	0.55150
FRR	0.25719	0.49647
CPINFND: BRR	0.57642	0.20486
FRR	0.73200	0.33671
CPIMMND: BRR	0.42258	0.20307
FRR	0.51441	0.16283

Source: Computed

[a]The CUSUM test has a 1 percent significance value of
1.143 and a 10 percent significance value of 0.850.

[b]The CUSUMSQ test for 32 observations and 4 parameters
has a 1 percent significance value of 0.3398 and a
10 percent significance value of 0.2301. For 23
observations and 5 parameters the 1 percent
significance value is 0.39922 and the 10 percent
significance value is 0.26794.

(Cont'd.)

^{c}The tests were performed on transformed data, using
the values shown in Table 2.
^{d}FFR = Forward recursive regression results;
 BBR = Backward recursive regression results.

five percent confidence level. These conclusions apply
to both the first and the second set of the estimated
price equations.

DETERMINANTS OF QUARTERLY COMMODITY PRICE MOVEMENTS:
MODEL SPECIFICATION AND EMPIRICAL ESTIMATES

In the short term commodity prices are influenced
by factors whose significance tends to get blurred in
the analysis of annual changes, notably the interaction
of "speculative" (non-industrial) demand and
"transaction" (industrial) demand. Developments
related to the physical availability of commodities--
strikes, transport problems, etc.--also plays a role.
If it is impossible to account for impact of these
latter factors satisfactorily, the influence of the
former can at least in part be taken into account by
directly incorporating in the demand equation some of
the variables that are a proxy for the behavior of
"speculators".

The basic framework illustrated in the previous
section needs additional modifications to be applied to
the analysis of short term aggregate price
relationships. While it is probably quite realistic
to maintain the assumption of competitive markets,
that of instantaneous adjustment of prices to demand-
supply changes might not hold in the short-term. It is
necessary to account in some way for the possibility
of adjustment costs in going from actual to equilibrium
prices. This can be done by postulating a "partial
adjustment" mechanism for the explanation of price
changes and by building it into the reduced form of the
model, which becomes:

$$P_t = \alpha_0 + \alpha_1 P_{t-1} + \alpha_2 IPIOE_t + \alpha_3 ERG_{t-1}$$
$$+ \alpha_4 GEERCV_t + \alpha_5 TIME \tag{5}$$

where:
P = relative commodity price index $\dfrac{CPI}{AMPIG}$

IPIOE = index of OECD industrial production
ERG = eurodollar rate in London
GEERCV = coefficient of variation of US$/DM
         exchange rate
TIME = time trend

TABLE 5
Constancy Tests on the Estimated Coefficients of Price Equations from Moving Regressions

	Homogeneity Statistics	Moving Regression Length	F Distribution $(V_1$ and $V_2)$	Critical F Values (5% Significance)
1948-1980				
CPI30D	0.183	10	(8,20)	2.45
CPIFDND	0.030	12	(4,24)	2.78
CPINFND	0.452	16	(4,24)	2.78
CPIMMND	0.336	12	(4,24)	2.78
1957-1980				
CPI30D	0.020	11	(5,13)	3.02
CPIFDND	0.026	11	(5,13)	3.02
CPINFND	0.907	11	(5,13)	3.02
CPIMMND	0.281	10	(5,13)	3.02

Source: Computed

and α_0, α_1, α_2, α_3, α_4 and α_5 are the summary expressions of the parameters of the reduced form, in terms of those of the structural equations,[19] embodying the "partial adjustment" transformation.[20]

The coefficients of the industrial production (IPIOE) and of the exchange rate instability variables (CEERCV) are expected to have a positive sign. Instability in exchange markets, here represented by the coefficient of variation of the daily exchange rates between the U. S. dollar and the deutsch mark, is postulated to induce short-term "hedging" purchases of commodities, therefore, increasing their demand and prices. The coefficients of the time trend (TIME) and interest rate (ERG) variables are, on the contrary, expected to be negative. Increases in interest rates, here represented by the three month rate on eurodollar deposits in London, raise the opportunity costs of holding commodity stocks and should therefore have a negative impact on prices.

In estimating equation (5) on quarterly data it was considered preferable to enter the interest rate variable in terms of changes from the previous quarter and the exchange rate instability variable as an index of the values of the coefficient of variation within each quarter. Only the price and activity variables (P and IPIOE) were entered as deviations from their respective time trends. Time was, therefore, not entered explicitly as an independent variable in the estimated equations.

The results of the econometric estimates of the quarterly price model applied to aggregate index (DEV3OQ), food (DEVFQ), non-food (DEVNFQ) and metals (DEVMQ) are shown in Tables 6, 7, 8 and 9 respectively. For each of the commodity groups, as well as for their aggregate, the econometric estimates of the price model contain three alternatives in addition to the basic version. Each of the three alternative equations for the aggregate price index, food, non-food and metal prices included a different dummy variable for the 1973-IV and 1974-I to 1974-IV quarters (DMOIL1, DMOIL2, DMOIL3). The three alternate equations for food also include a dummy variable for the Brazilian coffee frost, whose major effects on all tropical beverage prices were felt in the first two quarters of 1977 (DMCOF).

The three different OIL dummy variables were designed to test systematically the impact of the special events that occurred from the fourth quarter of 1973 to the fourth quarter of 1974 and to differentiate between the increase in oil and food prices. DMOIL1 is set equal to 1 from 1973-IV to 1974-IV and to zero in all other quarters. It represents the "strong" version of the oil crisis hypothesis: a lasting effect of the oil crisis on commodity prices through 1974. DMOIL2 is set equal to

TABLE 6
Relative Aggregate Commodity Price Equations (Quarterly Data): 1971-80[a]

Dependent Variable		Independent Variables			
	Constant	DEVIPIQ	ERG(-1)	GEERCV	DEV30Q(-1)
(1) DEV30Q	-3.18 (-1.63)	1.29* (4.31)	-0.13 (-1.71)	2.11* (2.31)	0.68* (10.04)
(2) DEV30Q	-3.97 (-1.97)	1.26* (4.23)	-0.13 (-1.76)	0.59* (5.91)	
(3) DEV30Q	-3.04 (-1.62)	1.17* (3.91)	-0.11 (-1.48)	1.76 (1.94)	0.64* (9.24)
(4) DEV30Q	-3.17 (-1.57)	1.29* (4.01)	-0.13 (-1.51)	2.11* (2.27)	0.69* (8.24)

Cont'd...

TABLE 6 cont'd.

Dependent Variable	Independent Variables			Test Statistics		
	DMOIL1	DMOIL2	DMOIL3	Adj R^2	D-W	S.E.E.
(1) DEV3OQ				0.86	1.77	6.18
(2) DEV3OQ	6.27 (1.32)			0.85	1.57	6.11
(3) DEV3OQ		9.55 (1.83)		0.86	1.79	5.97
(4) DEV3OQ			-0.06 (-0.01)	0.85	1.77	6.27

Source: Computed

[a] Estimated by OLS. (t values in parentheses).

* Signifciant at the 95% confidence level or above.

TABLE 7
Relative Food Commodity Price Equations (Quarterly Data): 1971-80[a]

Dependent Variable	Constant	Independent Variables			
		DEVIPIQ	ERG(-1)	GEERCV	DEVFQ(-1)
(1) DEVFQ	-5.67 (-1.95)	1.15* (2.89)	-0.06 (-0.60)	2.97* (2.22)	0.73* (9.52)
(2) DEVFQ	-8.56 (-3.17)	0.70 (1.87)	-0.07 (-0.73)	3.19* (2.71)	0.51* (5.49)
(3) DEVFQ)	-5.59 (-1.92)	1.03* (2.44)	-0.05 (-0.46)	2.73 (2.00)	0.71* (8.86)
(4) DEVFQ)	-5.82 (-1.93)	1.10* (2.39)	-0.05 (-0.43)	2.98* (2.20)	0.72* (8.09)

Cont'd...

TABLE 7 cont'd.

Dependent Variable	Independent Variables				Test Statistics		
	DMOIL1	DMOIL2	DMOIL3	DMCOF	Adj R^2	D-W	S.E.E.
(1) DEVFQ				16.12* (2.47)	0.79	2.12	8.68
(2) DEVFQ	18.29* (5.62)			18.98* (3.27)	0.84	1.98	7.64
(3) DEVFQ		6.54 (0.87)		16.20* (2.48)	0.79	2.09	8.71
(4) DEVFQ			1.95 (0.23)	16.31* (2.45)	0.79	2.11	8.80

Source: Computed

aEstimated by OLS. (t values in parentheses).

*Significant at the 95% confidence level or above.

TABLE 8
Relative Non-Food Commodity Price Equations (Quarterly Data): 1971-80[a]

Dependent Variable	Independent Variables				
	Constant	DEVIPIQ	ERG(-1)	GEERCV	DEVNFQ(-1)
(1) DEVNFQ	0.62 (0.15)	2.39* (2.80)	-0.11 (-0.85)	0.39 (0.46)	0.31 (1.72)
(2) DEVNFQ	1.11 (0.31)	2.63* (3.10)	-0.11 (-0.72)	0.57 (0.68)	0.34 (1.92)
(3) DEVNFQ	1.13 (0.26)	2.35* (2.98)	-0.11 (-2.03)	-0.11 (-0.14)	0.28 (1.70)
(4) DEVNFQ	0.68 (0.17)	2.31* (2.54)	-0.12 (-1.75)	0.39 (0.46)	0.34 (1.58)

Cont'd...

TABLE 8 cont'd.

Dependent Variable	Independent Variables			ρ	Test Statistics		
	DMOIL1	DMOIL2	DMOIL3		Adj R^2	D-W	S.E.E.
(1) DEVNFQ				0.73	0.54	1.46	6.26
(2) DEVNFQ	-8.28 (-1.61)			0.68	0.59	1.66	6.13
(3) DEVNFQ		12.35* (4.70)		0.77	0.60	1.32	5.71
(4) DEVNFQ			-1.78 (-0.27)	0.72	0.53	1.49	6.35

Source: Computed

a Estimated by OLS, using the Cochrane-Orcut correction for first order serial correlation. (t values in parentheses).

* Significant at the 95% confidence level or above.

TABLE 9
Relative Metal Price Equations (Quarterly Data): 1971-80[a]

Dependent Variables	Independent Variables				
	Constant	DEVIPIQ	ERG(-1)	GEERCV	DEVMQ(-1)
(1) DEVMQ	-3.54 (-1.13)	1.99* (3.47)	-0.27* (-2.77)	2.38 (1.89)	0.53* (4.77)
(2) DEVMQ	-3.79 (-1.17)	2.01* (3.37)	-0.27* (-2.79)	2.03 (1.57)	0.43* (2.90)
(3) DEVMQ)	-3.61 (-1.48)	1.60* (3.92)	-0.24* (-2.59)	1.78 (1.50)	0.54* (6.60)
(4) DEVMQ	-4.26 (-1.38)	1.89* (3.28)	-0.10 (-1.12)	1.77 (1.63)	0.31* (2.55)

Cont'd...

TABLE 9 cont'd.

Dependent Variable	Independent Variables			ρ	Test Statistics		
	DMOLI1	DMOLI2	DMOLI3		Adj R^2	D-W	S.E.E.
(1) DEVMQ				0.35	0.65	1.87	8.44
(2) DEVMQ	7.29 (0.94)			0.39	0.62	1.81	8.46
(3) DEVMQ)		20.72* (3.05)		-	0.83	1.93	7.84
(4) DEVMQ			25.64* (3.36)	0.48	0.67	1.91	7.41

Source: Computed

[a]Estimated by OLS, using the Cochrane-Orcut correction for first order serial correlation.

(t value in parentheses).

* Significant at the 95% confidence level or above.

1 in 1973-IV and 1974-I and to zero in all other
quarters. It represents a weak version of the oil
crisis hypothesis: strong, but short-term impact on
commodity prices. DMOIL3 is variant of the previous
case. It is set equal to 1 in 1974-I and 1974-II and
to zero in all other quarters.

The results indicate that the business cycle
indicator--i.e. the trend deviations of industrial
production in OECD countries--is shown to have a
strongly positive and significant effect on relative
commodity prices. A one percent deviation of
industrial production from its trend generates, ceteris
paribus, a 1.17 percent deviation of the aggregate
relative price index from its trend in the short term
and a 3.25 percent deviation in the long term. As
expected, non-food industrial raw material and metal
prices, are more sensitive than food prices to changes
in industrial demand conditions. If industrial
activity deviates from its trend by one percent, the
average short term percentage deviation of prices from
trend is 2.35 for non-food, 1.60 for metals, as
opposed to 0.70 for food prices. The corresponding
long term price deviations from trend are 3.26 percent,
3.47 percent and 1.42 percent for non-food, metal and
food prices respectively.[21/]

The lagged (relative) price variable is also
highly significant in all cases except for non-food.
The coefficient values indicate that adjustment of
quarterly prices to equilibrium values takes place
gradually, if rather rapidly. The results obtained
indicate that the speed of adjustment is greater for
non-food than for food and metal prices: is equal
to 0.72 for non food prices, 0.49 for food prices and
0.46 for metal prices.[21/]

The interest rate variable shows the expected
negative sign in all equations and specifications, but
it is statistically significant at the 95 percent
confidence level or above only in the case of non-food
and metals. Considering the approximate nature of the
interest variable used in the equations, these results
are very encouraging; they indicate quite clearly
that in the short term, interest rate changes influence
relative commodity prices in the expected negative
direction and that their effect is stronger and more
significant for metals and non-food agricultural raw
materials than for food products.

Exchange rate instability also influences relative
commodity prices in the expected positive direction.
The proxy used here for exchange rate instability is
found to have the right sign in all price equations
and almost all specifications, and to be statistically
significant at the 95 percent confidence level or
above for the aggregate of prices and food and at the
90 percent confidence level in the case of metal

prices. The greater is the instability in foreign exchange markets, the stronger the hedging demand for commodities seems to become. Exchange rate volatility apparently creates an impulse for holders of liquid assets to spread their short term holdings not only across currencies, but also across different types of assets, including commodities.

Among commodity groups, the effect of exchange rate instability seems to be stronger and more significant in the case of food and metal prices than in the case of non-food prices. The results concerning food prices are somewhat surprising, since one would not expect exchange rate hedging to create a stronger short-term demand for food than for other commodities. They may be the consequence of the partial failure of the exchange rate instability variable to differentiate between trend movements in the value of the U. S. dollar (the numeraire currency of the price indexes) and the fluctuations around this trend.

Finally, the quarterly model results seem to conform fully to those of the annual model with respect to the impact of the so-called oil crisis of 1973 on relative commodity prices. DMOIL1 has the expected positive sign and is statistically significant for food, but not for metal prices. In the case of non-food prices this variable even exhibits the wrong sign. This indicates that DMOIL1 picks up the effect of the cereal crisis and not those of the oil price increase of 1973. Conversely, DMOIL2 is shown to have the correct sign and to be significant for both metal and non-food prices, but not for food prices. The "oil crosis" seems to have had per se a limited effect on relative commodity prices: it apparently affected only the markets for metals and non-food products and for a relatively short time (no more than two quarters). The coffee frost in Brazil, proxied by DMCOF, on the contrary, appears to have had by itself a strong impact on the index of food prices, because of coffee's huge weight and the indirect effect of the coffee price rise in those of the other tropical beverages: almost half that of the "cereal crisis" of 1972-74.

CONCLUSIONS

Although this paper can only be considered a first step, it shows the importance of integrating "real" and "monetary" determinants in the analysis of commodity price movements.

Annual changes in non-oil commodity prices relative to those of manufactured products, that is in the net barter terms of trade between these primary commodities and manufactures, are shown to depend on the business cycle in industrial countries, on the growth of the international monetary base, and on special factors

affecting the supply of some of the more important
commodities. Aside from the shorter-term business
cycle in industrial countries and special supply
factors, quarterly commodity price changes are shown to
be influenced by instability in foreign exchange
markets and interest rate variations. The impact of
some of the "monetary" factors affecting the total
demand for commodities (transaction and speculative)
varies among commodity groups, but seems to be
sufficiently strong to carry over to the aggregate
price indexes as well.

The hypothesis of an asymmetrical impact of the
economic cycle on relative commodity prices (stronger
on its up than on its down side) does not find general
support in our analysis. Some downward stickiness of
prices is evident in the case of non-food products.
Neither do we find general support for the sub-
hypothesis that asymmetry in the price response to
demand conditions might have increased in the 1970s,
with respect to the previous decades. Developing
exporting countries do not seem to have developed any
great extra degree of "commodity power" in recent years.

Again, contrary to some "current wisdom" and at
variance with some recent empirical evidence (OECD),
our analysis also tends to reject the hypothesis that
there have been substantial structural changes in the
mechamisms of non-oil primary commodity price
determination in the post-war period.

NOTES

1. Non-oil is here used to mean non-oil exporting
developing countries. The NICs are the newly
industrializing developing countries.

2. Measured by our index of 30 commodity prices
traded in international markets which includes 18
agricultural food products (coffee, cocoa, tea, sugar,
beef, bananas, oranges, maize, rice, sorghum, wheat,
palm oil, coconut oil, groundnut oil, soybeans, copra,
groundnuts and soybean meal), six agricultural non-food
products (cotton, jute, rubber, wool, timber and hides
and skins) and six metals (copper, tin, aluminum,
nickel, lead and zinc). This is base weighted index,
reflecting 1975 weights, which represent the relative
value of each commodity in world trade. Average
monthly quotations in U. S. dollars of these
commodities on world markets were used to construcct
the index. (All data are available from the authors
on request.)

3. Measured by the UN index of manufactured
product prices (SITC 5-8) exported by industrial
countries.

4. For a recent analysis of long-term trends in
"real" commodity prices spanning from the 1860s to the

1970s see Ray (1977). Spraos (1980) carefully reviews
the debate concerning the terms of trade between
primary commodities and manufactures over the long
term.
    5. The changes in industrial production in OECD
countries are a good proxy for demand conditions in
commodity markets, as metals and non-food agricultural
raw materials are inputs into industrial processes and
some of the agricultural food products undergo
substantial transformations before being consumed.
OECD countries represent the major source of import
demand for most commodities (particularly metals, non-
food agricultural raw materials, tropical beverages and
vegetable oils).
    6. The measure of potential output in the major
industrial countries is that of Artus (1977).
    7. As a footnote to "history", these last two
studies originated in the World Bank and IMF in 1974-75,
when a lively debate sprang up between the
"structuralists", who believed that the changes in
relative commodity prices which occurred in 1973-74 had
permanently changed the long-term trends and the
"conjuncturalists" who viewed the 1973-74 price boom as
the result of cyclical circumstances, reversible in the
short term. The controversy was not resolved at the
time--and perhaps rightly so, as the evidence then
brought to bear was far from conclusive on either side
--but the actual market performance of commodity prices
in the second half of the 1970s strongly supported the
view that the 1973-74 price boom was cyclical and
reversible.
    8. The only exceptions are metal prices that
experienced a somewhat greater instability in the 1960s
than in the 1970s.
    9. The UN index of unit values of manufactured
products that are internationally traded (in U. S.
dollars) is used here as a proxy for AMPIG.
    10. As a variant, gold holdings by monetary
authorities were added to it (both at official prices
and at market prices).
    11. It is also possible to postulate a positive
relationship between changes in this ratio and
speculative purchases in commodity markets. What we
try to measure here, however, is not the strength of
speculative purchases and sales of commodities, but the
more permanent trend in the substitution of commodity
assets for liquid assets.
    12. For each equation, the 1957-80 estimates are
presented with and without the ILR variable.
    13. Purchases of metals by holders of liquid
assets denominated in dollars for hedging against
changes in the value of the US currency can also be
postulated. These, however, would be short-term in

nature and should not be strongly reflected in annual price observations.

14. Oil related expectations might have generated some panic and/or speculative buying of commodities in 1974. Panic purchases were widely reported in 1974 (especially, but not exclusively, from Japan). We simply mean that the impact of these phenomena does not appear to have lasted long when the industrial outlook in OECD countries weakened.

15. According to FAO data, world stocks of cereals which were at 168 million metric tons at the end of the 1971/72 season, fell to 119 million metric tons at the end of the 1972/73 season and went further down to 108 million metric tons in the course of the 1973/74 season. (FAO, The State of Food and Agriculture, 1978, Rome, 1979.)

16. Some shortages of fertilizers in developing importing countries were also reported.

17. It is known, however, that both tests are approximate and that the CUSUMSQ test, in particular, tends to give significant results (and to lead to a rejection of the null hypothesis) more often than an exact test would for small samples. Only when the number of observations minus the number of the parameters of the estimated equations is greater than 30, does the discrepancy between approximate and exact test results diminish.

18. The difference between the CUSUM and CUSUMSQ tests lies in the fact that the former catches the systematic instability in the regression coefficients, while the latter also captures the haphazard deviations of the estimated coefficients from a constant value, since it disregards the sign of the forecast errors.

19. These are: a) $Qd_t = a_0 + a_1 P_t + a_2 IPIOE_t$

$$+ a_3 ERG_{t-1} + GEERCV$$

b) $Q_{st} = b_0 + b_1 P_t + O_{b2} TIME$

20. $P_t - P_{t-1} = \delta (P_t^* - P_{t-1})$

where: $0 < \delta < 1$

21. These results were derived from the "preferred" estimates: equation (3) in Tables 6, 8 and 9 and equation (2) in Table 7.

REFERENCES

Artus, J. "Measures of Potential Output in Eight Industrial Countries, 1955-78", IMF Staff Papers. 1 (1977): 1-35.

Biasco, S. L'Inflazione nei Paesi Capitalistici Industrializzati Milano: Feltrinelli Editore, 1979:

91-100.

Brown, R. L., J. Durbin, and J. M. Evans. "Techniques for Testing the Constancy Regression Relationships over Time". Journal of the Royal Statistical Society. Series B (1975): 149-192.

Chu, K. Y. "Short Run Forecasting of Commodity Prices: An Application of Autogregressive Moving Average Models". IMF Staff Papers. 1 (1978): 90-111.

Cooper, R. N. and R. Z. Lawrence. "The 1972-75 Commodity Boom". Brookings Papers on Economic Activity. 3 (1975): 671-715.

Enoch, C. A. and M. Panic. "Commodity Prices in the 1970s". Bank of England Quarterly Bulletin. 1 (1981): 42-53.

Fratianni,M., and P. Savona. La Liquidita Internazionale. Bologna: Il Mulino, 1972.

Fratianni, M., and P. Savona. "International Liquidity: An Analytical and Empirical Reinterpretation". Ente Einaudi Quaderni di Ricerche. 11 (1973): 47-121.

Goldstein, M. "Downward Price Inflexibility, Ratchet Effects and the Inflationary Impact of Import Price Changes: Some Empirical Evidence". IMF Staff Papers. 3 (1977): 569-612.

Goreux, L. M. "Commodity Prices and Business Cycle". IMF Research Department. draft mimeo, 1975.

Grilli, E. "Commodity Price Movements Before and After the Oil Crisis". The World Bank. draft mimeo, 1975.

Hwa, E. C. "Price Determination in Several International Primary Commodity Markets". IMF Staff Papers. 1 (1979): 157-188.

Labys, W. "Commodity Markets and Models: The Range of Experience". In Stabilizing World Commodity Markets: Analysis Practice and Experience, edited by F. G.Adams. Lexington, Mass.: Lexington Books, 1978.

OECD. Commodity Price Behavior. Doc. DES/WPI,EM(81)6 (restricted). Paris, 1981.

Radetzki, M. "Commodity Prices During Two Booms". draft mimeo, 1974.

Ray, G. F. "The 'Real' Price of Primary Products". National Institute Economic Review. 81 (1977): 72-76.

Ray, G. F. and H. J. Timm. "Forecasting Commodity Prices". National Institute Economic Review. 3 (1980): 76-79.

Richard, D. "A Dynamic Model of the World Copper Industry". IMF Staff Papers, 4 (1978): 779-833.

Spraos, J. "Statistical Debate on the Net Barter Terms of Trade Between Primary Commodities and Manufactures". The Economic Journal. 90 (1980): 107-128.

Wymer, C. R.   "Estimation of Continuous Time Models
    with an Application to the World Sugar Market".
    In Quantitative Models of Commodity Markets, edited
    by W. C. Labys.   Cambridge, Mass.:   Lexington Books,
    1975: 173-191.

2
Market Structure, Information, Futures Markets, and Price Formation

Albert S. Kyle

INTRODUCTION

The futures markets for agricultural and other
commodities are easy to single out as being those
capitalist institutions which come closest to
satisfying the textbook definition of perfect
competition. Indeed, the large open interest, heavy
trading volume, and great liquidity of the active
futures markets for soybeans, corn, and a variety of
other commodities seem to make the assumption of
perfect competition a reasonable one. This assumption
gains further plausibility when one considers that the
market clearing price is determined by the non-
cooperative actions of hundreds of floor traders and
thousands of speculators and hedgers off the floor of
the commodity exchanges.

Two facts about futures trading suggest, however,
that futures markets are not as competitive as they
seem at first glance. First, seats on commodity
exchanges are quite valuable and that much of their
value is derived from the floor trading privileges
attached to their ownership. The barrier to entry
created by the requirement that floor traders own
seats may well increase the returns to floor traders
by making floor trading less competitive.[1] Second,
open interest in futures markets is often concentrated
in the hands of a small number of traders. This
suggests that traders with large positions will
consider the effect of their trading actions on price
and thus behave in a strategic or, equivalently, non-
competitive manner.

In this paper, a simple model of the trading
process is constructed which incorporates the

Albert S. Kyle, Department of Economics, Princeton
University, Princeton, N. J.

45

assumption that commodity futures markets are non-competitive in a manner consistent with the above facts. The model is used to analyze the effectiveness of policies designed to improve the performance of agricultural commodity markets by increasing the informativeness of prices. In particular, the following questions are addressed:

1. How does the publication of information, such as crop forecasts and export data by the USDA, affect the informativeness of prices?

2. How does a policy of excluding from trading in commodity futures markets those "retail speculators" who have no economic need to trade in the market and no non-public information about demand or supply for the commodity tend to affect the amount of noise in prices and therefore the informativeness of prices?

These questions concerning the informativeness of speculative prices are clearly important ones. The USDA currently spends millions of dollars each year collecting and publicizing information about atricultural commodities; yet economists have not been able to come to grips with the question of what would happen if the production of information were left to "market forces", especially if these market forces are not competitive. Concerning trading by uninformed retail speculators, the Commodity Futures Trading Commission, which regulates organized trading in agricultural and other commodities on commodity exchanges, has a policy of allowing essentially anyone who wants to trade futures contracts to do so as long as he has a sufficiently large amount of money to lose. While there are virtually no suitability rules to weed out uninformed or emotionally ill-equipped traders, it has been proposed that such rules would not only be desirable on paternalistic grounds but would also tend to stabilize prices by eliminating uninformed noise trading. This problem is also one which economists have a hard time addressing, especially with imperfectly competitive markets.

The model of imperfectly competitive speculative markets discussed in this paper has properties quite different from models based on the assumption of perfect competition. The assumption that floor traders, who provide a market making service, behave as imperfect competitors leads to negative serial correlation in price movements. This negative serial correlation allows market makers to make supernormal profits by buying when other traders are selling (and prices are falling) and selling when other traders are buying (and prices are rising). In addition to imperfectly competitive market makers, the model has large imperfectly competitive speculators who trade on the basis of private information. Since these large speculators trade strategically, taking into account the effect that their buying and selling has on market prices, they find it optimal to trade smaller quantities than might otherwise be optimal in order to obtain better prices. This restricted trading means that large speculators in effect withhold some of their private information from prices. In equilibrium, prices do not reveal all of the private information of large speculators but large speculators earn positive profits on the basis of their private information.

In competitive models, the analogous features the equilibrium depend very heavily upon the assumption of risk aversion. In our model of imperfectly competitive speculative markets, both market makers and speculators are risk neutral. In competitive markets, however, risk aversion is necessary if market makers or speculators are to earn positive returns. With perfectly competitive risk averse market makers, prices have negative serial correlation due to the fact that when prices are falling (because other traders are selling) market makers are unwilling to buy at actuarially fair prices because of risk aversion; a similar argument holds when prices are rising. When competitive speculators with private information are risk averse, they are unwilling to trade large quantities at the statistically unbiased prices which reveal all of their private information because these prices offer no return for bearing risk. In equilibrium, the risk premium speculators demand prevents prices from revealing all of the private information of speculators.

The assumption that speculative markets are imperfectly competitive has policy implications which differ from those of standard competitive models. Consider first the effect of the public production of new information which is an imperfect substitute for speculators' private information. We show below that an increase in public information tends to make the atmosphere for speculators more competitive and thus induces them to incorporate more of their private information into prices. This lowers the returns to speculators and leads to an exit of speculators from the market, but prices in the new equilibrium are more informative than prices were before the increase in public information.

The greatest difference between imperfectly competitive and competitive models concerns the effect of noise trading on the informativeness of prices. In competitive models, increases in noise trading lead to increases in the size of risky positions that must be carried by speculators and market makers. The

increased risk premiums demanded by speculators and
market makers to carry these positions make prices
noisier and therefore less informative. While returns
to speculators and market makers increase and new entry
occurs, prices in the new equilibrium which emerges are
no more informative than in the old. In this sense,
uninformed noise trading is destabilizing.

In our model of imperfectly competitive markets,
by contrast, increased noise trading tends to make
prices more informative. This surprising result is due
to the tendency for increased noise trading to increase
the liquidity of the market, where "liquidity" is
proportional to the quantity of trade it takes to cause
prices to change by one dollar. When noise trading
increases, market liquidity increases and speculators
and market makers, acting similarly to monopolists who
suddenly face a more elastic demand curve, are willing
to trade the larger quantities necessary to sustain the
greater degree of liquidity, with prices equally as
informative as before. In this situation the rewards
to speculators are increased, new entry occurs, and in
the new equilibrium which emerges, prices are more
informative because more information is produced and
more of the information which is produced is
incorporated into prices.

The obvious policy implication of this result is
that it is not clear that uninformed noise trading is
bad for anyone other than the noise traders themselves.
The case for suitability rules must be based upon
specific assumptions about the structure of speculative
markets or upon paternalistic arguments.

HOLBROOK WORKING'S IDEAS AND THE RECENT LITERATURE

This paper draws upon ideas from four strands of
literature in finance and economics: the pioneering
work of Holbrook Working on futures markets, the
theoretical and empirical work on the Efficient Markets
Hypothesis, the rational expectations literature, and
the industrial organization literature.

Many modern ideas about the informational role of
prices were formulated quite early by Working (1953).
Working observed if prices reflect expectations
accurately, then price fluctuations in an "ideal"
market are unpredictable, reflecting always the arrival
of new information. He also formulated clearly the
notion that if futures prices were always the best
possible estimate of prices at a later date, then
returns to speculators would be driven to zero and "the
speculation necessary to maintain even an approximation
to ideal price behavior would tend to vanish." He
proposed that economists investigate price fluctuations
empirically in an effort to isolate the "imperfections"

which enable professional traders to make profits
consistently. On the basis of his own rudimentary
empirical work (1953, 1960, 1967) Working observed that
futures prices tend to jiggle back and forth on an
intra-day, tick-by-tick basis, and the resulting
negative autocorrelation in price changes enables floor
traders who make markets (scalpers or day traders) to
earn positive profits consistently; he also observed
that it is very difficult to construct, on the basis
of current public information, better estimates of the
future price of wheat than its current price.

In the efficient markets literature of the 1960s
and 1970s we find a more sophisticated approach to the
theoretical assumptions underlying the Efficient
Markets Hypothesis and a great deal of empirical work
of the kind advocated by Working (much of the empirical
work being done on stock market data rather than
futures market data). Fama (1970) provides a well-
known summary of the earlier empirical work in the area,
together with a treatment of three distinct forms of
the Efficient Markets Hypothesis: the weak form, the
semi-strong form, and the strong form. Fama's
conclusions about the efficiency of stock market prices
are very similar to Working's conclusions about futures
market data. It is difficult to beat the market on the
basis of public information. (The stock market is semi
-strong form efficient.) Only two glaring
inefficiencies stand out: a tendency towards too many
reversals of tick-by-tick prices, documented by
Niederhoffer and Osborne (1966), reflecting the
supernormal profits earned by specialists due to their
monopolistic access to the order flow (consistent with
Working's evidence on day trading); and an ability of
corporate insiders to earn supernormal profits on the
basis of inside information which is not
instantaneously incorporated into prices.

The theoretical model discussed in this paper
captures exactly these kinds of informational
imperfections while preserving the assumptions that it
is impossible to make profits consistently on the
basis of public information. Furthermore, Fama's
assertion that the profit opportunities are due to
monopolistic access to information is exactly the idea
designed to be captured by our model. Thus, our model
fits well empirical evidence from the efficient markets
literature, assuming of course that futures markets for
agricultural commodities behave something like markets
for common stocks.

A theoretical tool useful in our model is the
concept of a rational expectations equilibrium, due to
Muth (1961), in which a sharp distinction is made
between changes in structure and changes in outcomes.
As in a rational expectations equilibrium, our traders

are assumed to know the structure of the model when making trading decisions but not the outcome of all random variables they would like to know. Our result that uninformed trading is stabilizing is a <u>structural</u> result. That is an increase in the variance of the quantity of noise trading leads to an increase in the statistical precision of prices. At the same time, a large <u>outcome</u> of the quantity of noise trading, which traders do not observe is associated with a noisier, less precise price.

Our model also deals with the role of noise trading in sustaining a rational expectations equilibrium, an issue discussed for example by Grossman and Stiglitz (1980). In the context of a formal model of a competitive market with informed and uninformed traders, Grossman and Stiglitz show that as trading noise vanishes, prices become perfectly informative. In the limit, equilibrium breaks down because informed traders cannot earn a positive return on their private information. In our model (without entry and exit of speculators) the informativeness of prices is not affected by changes in the amount of noise trading. When entry and exit of speculators occurs, a reduction in noise trading leads to exit of speculators until eventually only one monopolistic speculator is left; even he eventually leaves as well. Prices become less (not more) informative with fewer speculators and although eventually there are no speculators and therefore, no private information incorporated into price, equilibrium does not break down. The introduction of non-competitive behavior on the part of informed traders thus eliminates the tendency for equilibrium to break down (with endogenous entry), as it does in the Grossman and Stiglitz model.

Our concept of imperfect competition is similar to concepts in the industrial organization literature. We use a Nash equilibrium in trading strategies for both market makers and speculators, with particular restrictions to capture the informational structure of the market discussed below.

Our "paradoxical" result that noise trading is stabilizing is similar to the result that an increase in demand may <u>lower</u> price in a model of Cournot competition. Consider a market with a linear demand curve and firms characterized by a fixed cost and a constant marginal cost the same for all firms. There is no competitive equilibrium because marginal cost pricing does not cover fixed costs, but with freedom of entry, quantity Cournot pricing leads to an equilibrium price above marginal cost. Now suppose that demand doubles. It can be shown that this leads to additional entry and a <u>lower</u> price because the market becomes more competitive. This result is similar to our result that more noise leads to more entry and a more competitive, informative price. The similarity is not a perfect analogy, however.

AN ECONOMIC FRAMEWORK

In this section we outline a simple economic framework within which the informativeness of prices can be discussed and from which implications about efficiency of resource allocation can be drawn. The model is a very simple one designed to make a simple economic point, but the idea can be generalized.

Consider a seasonal agricultural commodity which can be stored costlessly for two periods after the harvest but which perishes before the next harvest. The harvest, denoted z, becomes an "exhaustible resource" which is consumed over two periods. Letting c_1 and c_2 denote consumption in the two periods, we have the resource constaint $z = c_1 + c_2$. 2/

Demand for the commodity is generated by a representative consumer's quadratic utility function:

$$u(x, c_1, c_2) = x - \frac{1}{2}(A - c_1)^2 - \frac{1}{2}(A + \tilde{d} - c_2)^2, \quad (1)$$

where x denotes consumption of the numeraire good, A is a parameter defining the utility function, and $\tilde{d}$ is a random demand shock with zero mean affecting demand in period two.

Demand functions for the good in the two periods are given by:

$$c_1 = A - p_1, \quad (2)$$

$$c_2 = A + \tilde{d} - p_2, \text{ and}$$

where p_1 and p_2 denote prices in periods one and two respectively. To abstract away from unnecessary complications, we assume that storage costs and interest rates are zero. This allows us to interpret p_1 as a futures price for delivery in period two as well as a spot price.

If the outcome of $\tilde{d}$ is observed before the consumption decision must be made in period one, then storage arbitrage equates prices across periods, and equilibrium prices and quantities are given by:

$$p_1 = p_2 = A + \frac{1}{2}\tilde{d} - \frac{1}{2}z, \quad (3)$$

$$x_1 = \frac{1}{2}(z - \tilde{d}), \text{ and}$$

$$x_2 = \frac{1}{2}(z + \tilde{d}).$$

This allocation maximizes social welfare defined using the standard consumer surplus concept. We shall assume, however, that the outcome of $\tilde{d}$ is not observed with perfect accuracy in period one but that some information about the outcome of $\tilde{d}$ is incorporated into prices in period one. In this case actual social surplus will be lower by the amount $\frac{1}{4}(p_2 - p_1)^2$, where p_2, obtained from the demand function and the resource constraint, is given by:

$$p_2 = 2A + \tilde{d} - z - p_1. \tag{4}$$

We wish to develop the intuitive idea that an accurate price p_1 implies that $p_2 - p_1$ is small, thus establishing a positive relationship between the informativeness of prices, the stability of prices, and social welfare. To do so, assume that there exist risk neutral traders and that an ex ante market is held in which there is no information available about the outcome of $\tilde{d}$. Letting p_0 denote the ex ante price, it is easy to show that a policy of buying two units ex ante and selling one unit in periods one and two respectively yields profits $2p_0 - 2A + z + \tilde{d}$, an expression which does not depend on the particular prices p_1 and p_2 which are realized. Then the assumptions of risk neutrality (implying zero expected profits) and no information about $\tilde{d}$ (which has a zero "prior" mean) guarantee that $p_0 = A - \frac{1}{2}z$. Consider the price fluctuations $\Delta\tilde{p}_1$ and $\Delta\tilde{p}_2$ defined by $\Delta\tilde{p}_1$, $= \tilde{p}_1 - \tilde{p}_0$, $\Delta\tilde{p}_2 = \tilde{p}_2 - \tilde{p}_1$, where we now think of $\tilde{p}_1$ and $\tilde{p}_2$ as random variables. From equation (4), it is easy to show that:

$$\tilde{d} = 2\Delta\tilde{p}_1 + \Delta\tilde{p}_2. \tag{5}$$

Now make the Martingale assumption that $E(\Delta\tilde{p}_2 | \Delta\tilde{p}_1) = 0$. This allows us to interpret $2\Delta\tilde{p}_1$ as an unbiased forecast of $\tilde{d}$ with error $\Delta\tilde{p}_2$. The overall volatility of prices, measured by $\text{var}(\tilde{p}_2) = \text{var}(\Delta\tilde{p}_1) + \text{var}(\Delta\tilde{p}_2)$, can be shown to be given by:

$$\text{var}(\Delta\tilde{p}_2) = \text{var}(\tilde{d}) - 3\text{var}(\Delta\tilde{p}_1). \tag{6}$$

Improving the informativeness of the price $\tilde{p}_1$ increases the variance of $\Delta\tilde{p}_1$, but reduces the variance of $\Delta\tilde{p}_2$ four times as much; it thus reduces the overall variance of prices. Note that it also reduces the expected welfare loss due to not having perfect information, which is given by $\frac{1}{4}\text{var}(\Delta\tilde{p}_2)$.

The idea that more informative prices are less volatile overall and result in expected welfare gains is contained in Working's onion study (1960). He observed in that study that an increase in the informativeness of prices should increase volatility early in the season but reduce it by even more later on. His study sought to document this fact empirically using the historical experience of onion futures trading.

INFORMATION STRUCTURE AND TRADING STRATEGIES

While a framework for discussing the implications of prices is outlined in the previous section, that section does not contain a model of the price formation process itself. In principle, a large number of specific models of price formation, both competitive and non-competitive, can be developed within this framework. In this section we outline one such model of market structure and price formation.

Of the three prices p_0, p_1, and p_2, clearly p_1 is the only one which is determined in a non-trivial manner. The price p_0 is merely an <u>ex ante</u> expectation, and p_2 is determined mechanically from p_1 by equation (4). In our model of price formation, the price p_1 is the outcome of a game played by several different classes of traders. The rules of this game determine the institutional structure of the market. In the rest of this section we discuss "the rules of the game" by defining the trading strategies each trader can use, paying particular attention to the information each trader utilizes in his trading strategy.

There are three kinds of traders in the market: an undetermined number of uninformed noise traders, M market makers, and N informed speculators. These traders trade on the basis of public and private information about the demand shock $\tilde{d}$.

The public information, which can be interpreted as a government statistic published to all traders in advance of trading in period one, is a noisy observation of $\tilde{d}$ given by $\tilde{d} + \tilde{g}$, where $\tilde{g}$ is a random noise term. In addition to this public information, each of the N informed speculators, indexed $n = 1,...N$, has a private observation given by $\tilde{d} + \tilde{e}_n$, where $\tilde{e}_n$ is a random noise term. We assume that the N + 2 random variables $\tilde{d}$, $\tilde{g}$, $\tilde{e}_1,...,\tilde{e}_N$ are normally and independently distributed with zero means and variances given by:

$$\text{var}(\tilde{d}) = 1/\tau_0,$$

$$\text{var}(\tilde{g}) = 1/\tau_G,$$

$$\text{var}(\tilde{e}_n) = 1/\tau_I, \quad n = 1,...,N, \text{ and}$$

where the constants τ_0, τ_G, and τ_I are the precisions corresponding to the variances.

Utilizing the government statistic $\tilde{d} + \tilde{g}$, it is possible to mimic the procedure in Section 2 and calculate a revised ex ante price p_0^* defined by:

$$\tilde{p}_0^* = A - \frac{1}{2} z + \frac{1}{2} E(\tilde{d}|\tilde{d} + \tilde{g}), \tag{8}$$

where

$$E(\tilde{d}|\tilde{d} + \tilde{g}) = \frac{\tau_G}{\tau_0 + \tau_G} (\tilde{d} + \tilde{g}), \tag{9}$$

and $\quad \mathrm{var}^{-1}(\tilde{d}|\tilde{d} + \tilde{g}) = \tau_0 + \tau_G. \tag{10}$

We can interpret the difference $p_0^* - p_0 = \frac{1}{2} E(\tilde{d}|\tilde{d} + \tilde{g})$ as an "announcement effect" which increases the prior precision of p_0 from τ_0 to $\tau_0 + \tau_G$, i.e. adds τ_G units of precision to the ex ante price.

The trading strategy of each of the M market makers, indexed $m = 1,\ldots,M$, is a constant γ_m which determines the quantity the mth market maker buys, denoted $\tilde{x}_{Mm}$, according to the formula:

$$\tilde{x}_{Mm} = -\gamma_m (\tilde{p}_1 - \tilde{p}_0^*) \tag{11}$$

The quantity which $-\gamma_m$ multiplies on the right-hand-side is the unexpected fluctuation in prices. When γ_m is positive, market makers lean against the wind, buying when prices fall unexpectedly and selling when prices rise unexpectedly.

The strategy choice of each of the N speculators, indexed $n = 1,\ldots,N$, is a constant β_n which determines the quantity the nth informed speculator purchases, denoted $\tilde{x}_{In}$, according to the formula:

$$\tilde{x}_{In} = \beta_n [E(\tilde{d}|\tilde{d} + \tilde{g}, \tilde{d} + \tilde{e}_n) - E(\tilde{d}|\tilde{d} + \tilde{g})]. \tag{12}$$

The quantity which β_n multiplies on the right-hand-side of the above equation is the difference between the informed speculator's best forecast of the demand shock $\tilde{d}$ on the basis of public and own private information and the best forecast made on the basis of public information alone. Intuitively it measures what he knows that everyone else does not know. Denoting this quantity i_n, we have $x_{In} = \beta_n i_n$ and can calculate explicitly:

$$i_n = E(\tilde{d}|\tilde{d} + \tilde{g}, \tilde{d} + \tilde{e}_n) - E(\tilde{d}|\tilde{d} + \tilde{g}) \tag{13}$$

$$= \frac{\tau_I}{\tau_0 + \tau_G + \tau_I} [(\tilde{d} + \tilde{e}_n) - \frac{\tau_G}{\tau_0 + \tau_G} (\tilde{d} + \tilde{g})].$$

The uninformed noise traders trade "exogenously," purchasing a random quantity denoted $\tilde{u}$ in period one. The random variable $\tilde{u}$ is normally distributed with zero mean and variance σ_u^2; it is exogenous in that it is assumed to be distributed independently of $\tilde{d}$, $\tilde{g}$, $\tilde{e}_1,\ldots,$ $\tilde{e}_N$. Noise trading thus contains no information about "the fundamentals" of supply and demand.

Some discussion of the trading strategies of market makers and informed speculators is in order at this point. It is useful to think of the trading strategies as functions of the informational and price variables $\tilde{d} + \tilde{g}$, $\tilde{d} + \tilde{e}_n$, and $\tilde{p}_1$. Under this interpretation, the parameters γ_m and β_n represent particular linear functions of subsets of these variables. Using the concept of equilibrium defined below, it can be shown that linear strategies of the particular form specified by equations (11) and (12) are equilibrium strategies when market makers and informed traders choose from the following broader sets of strategies:

1. The mth market chooses the quantity he trades $\tilde{x}_{Mm}$ as any measurable function of $\tilde{d} + \tilde{g}$ and $\tilde{p}_1$ (but not of private observations $\tilde{d} + \tilde{e}_n$);

2. The nth informed trader chooses the quantity he trades $\tilde{x}_{In}$ as any measurable function of $\tilde{d} + \tilde{g}$ and $\tilde{d} + \tilde{e}_n$ (but not of the market price $\tilde{p}_1$). Proving this result is somewhat tedious and takes us beyond the scope of this paper.[3/] Hence, we have begun by assuming that the trading strategies have this particular linear form.

These trading strategies do, however, embody one restrictive assumption. While informed speculators can choose the quantities they wish to trade on the basis of public and private information, they must choose this quantity before they observe the market clearing price p_1. It is as if they are forced by the rules of the game to trade by placing a market order for a specific quantity. This assumption confers an informational advantage upon market makers by giving them a chance to react to the incoming order flow before informed speculators are given a chance to react. If we think of market makers as floor traders and informed speculators (and noise traders) as traders off the floor, then this assumption captures in a rough way an informational division of labor, which casual observation suggests is a characteristic of organized commodity trading. Floor traders concern themselves a great deal with the order flow but are often quite ignorant of information about the fundamentals of demand and supply; at the same time, even large

informed speculators off the floor find it impossible
to beat floor traders at their game. We have tried
to capture this institutional feature of commodity
trading with the assumption that market makers set
quantities traded as functions of prices but
speculators do not.[4]

THE DEFINITION OF EQUILIBRIUM AND ITS EXISTENCE

We now turn to the definition of equilibrium.
Roughly speaking, our equilibrium is a Nash equilibrium
in the parameters γ_m and β_n, calculated under the
assumption that market makers and informed speculators
are risk neutral. According to the Nash equilibrium
concept, each market maker and each informed trader
maximize expected profits taking into account his
effect on prices.

From the market clearing condition:

$$\sum_{m=1}^{M} \tilde{x}_{Mm} + \sum_{n=1}^{N} \tilde{x}_{In} + \tilde{u} = 0, \tag{14}$$

and from equations (11) and (12), the price in period
one is given by:

$$p_1 = p_0^* + \left(\sum_{m=1}^{M} \gamma_m \right)^{-1} \left(\sum_{n=1}^{N} \beta_n \tilde{i}_n + \tilde{u} \right), \tag{15}$$

and from equation (4) the price in period two is given
by:

$$\tilde{p}_2 = \tilde{p}_0^* + \tilde{d} - E(\tilde{d}|\tilde{d} + \tilde{g}) - \left(\sum_{m=1}^{M} \gamma_m \right)^{-1}$$
$$\left(\sum_{n=1}^{N} \beta_n \tilde{i}_n + \tilde{u} \right). \tag{16}$$

Note that the unexpected fluctuation in prices $\tilde{p}_1 - \tilde{p}_0^*$
is a linear function of the order flow, $\Sigma \beta_n \tilde{i}_n + \tilde{u}$, and
that this order flow term consists of "information plus
noise." In taking into account the effect that their
trading has on "prices," market makers and speculators
must consider their effect on both $\tilde{p}_1$ and $\tilde{p}_2$. Buying
more today raises $\tilde{p}_1$ but increases stocks and thus
lowers $\tilde{p}_2$ tomorrow. From equations (15) and (16), $\Delta \tilde{p}_2$
which represents profits per unit traded under the
assumption that positions are liquidated in period two,
is given by:

$$\Delta \tilde{p}_2 = \tilde{p}_2 - \tilde{p}_1 = \tilde{d} - E(\tilde{d}|\tilde{d} + \tilde{g}) - 2 \left(\sum_{m=1}^{M} \gamma_m \right)^{-1}$$
$$\left(\sum_{n=1}^{N} \beta_n \tilde{i}_n + \tilde{u} \right). \tag{17}$$

The expected profits of market makers and the expected profits of informed traders, denoted Π_m^M and Π_n^I, are given as functions of the strategy parameters $\gamma_1, \ldots, \gamma_M, \beta_1, \ldots, \beta_N$ by:

$$\Pi_m^M(\gamma_1, \ldots, \gamma_M, \beta_1, \ldots, \beta_N) = E\{-\gamma_m(\tilde{p}_1 - \tilde{p}_0^*)\Delta\tilde{p}_2\},$$

$$m = 1, \ldots, M, \qquad (18)$$

$$\Pi_n^I(\gamma_1, \ldots, \gamma_M, \beta_1, \ldots, \beta_N) = E\{\beta_n \tilde{i}_n \Delta\tilde{p}_2\},$$

$$n = 1, \ldots, N, \qquad (19)$$

where $\Delta\tilde{p}_2$ and $\tilde{p}_1$ are understood to depend upon $\gamma_1, \ldots, \gamma_M, \beta_1 \ldots, \beta_N$ according to equations (15) and (17). An <u>equilibrium</u> is defined as a vector of strategy parameters $\langle\gamma_1, \ldots, \gamma_M, \beta_1, \ldots, \beta_N\rangle$ such that γ_m maximizes $\Pi_m^M(\gamma_1, \ldots, \gamma_M, \beta_1, \ldots, \beta_N)$ holding constant the other $N + M - 1$ parameters and β_n maximizes $\Pi_n^I(\gamma_1, \ldots, \gamma_M, \beta_1, \ldots \beta_N)$ holding constant the other $N + M - 1$ parameters.

It can be shown that a unique equilibrium exists. The actual proof is tedious and takes us beyond the scope of this paper, but it goes as follows:[5] By substituting specific expressions for $\Delta\tilde{p}_2$ and $\tilde{p}_1$ into the expressions for $\Pi_m^M(\ldots)$ and $\Pi_n^I(\ldots)$ and evaluating the resulting expectations, specific expressions for Π_m^M and Π_n^I are calculated in terms of the strategy parameters and the six exogenous parameters M, N, τ_0, τ_G, τ_I and σ_u^2. These expressions are then differentiated with respect to the appropriate strategy parameters to obtain $N + M$ first order conditions characterizing the $N + M$ equilibrium strategy parameters in terms of the six exogenous parameters. A symmetry argument is used to show that all market makers choose the same strategy parameter and all informed speculators choose the same strategy parameters. Call these parameters γ and β respectively. This reduces the $N + M$ first order conditions in $N + M$ unknowns to two equations in two unknowns. These two equations are then solved for γ and β explicitly. The results of these calculations are the following messy-looking expressions:

$$\gamma = \frac{2(M-2)(N+1+\frac{2\tau_I}{\tau_0+\tau_G})(\sigma_u^2)^{1/2}}{M(M-1)\,N^{1/2}(\frac{1}{\tau_0+\tau_G} + \frac{1}{\tau_I})^{1/2}}, \text{ and} \qquad (20)$$

$$\beta = \frac{(M-2)}{(M-1)} \frac{(\sigma_u^2)^{1/2}}{N^{1/2}(\frac{1}{\tau_0+\tau_G} + \frac{1}{\tau_I})^{1/2}} \tag{21}$$

Now define λ by:

$$\lambda = \frac{1}{M\gamma}, \tag{22}$$

and let $\tilde{x}$ denote the combined quantity bought by speculators and uninformed traders:

$$\tilde{x} = \beta \sum_{n=1}^{N} \tilde{i}_n + \tilde{u}. \tag{23}$$

Then p_1 is given by:

$$\tilde{p}_1 = \tilde{p}_0^* + \lambda\tilde{x}. \tag{24}$$

The quantity λ, which converts order flow into price fluctuations, is a good index for the liquidity of the market, with small values of λ corresponding to great liquidity. It is not too misleading to refer to λ as the equilibrium bid-asked spread. The total price fluctuation $\tilde{p}_1 - \tilde{p}_0$ can be written:

$$\tilde{p}_1 - \tilde{p}_0 = (\tilde{p}_0^* - p_0) + \lambda\tilde{x}. \tag{25}$$

The term $\tilde{p}_0^* - p_0 = \frac{1}{2}E(\tilde{d}|\tilde{d} + \tilde{g})$ is an announcement effect and the term $\lambda \tilde{x}$ is an order flow effect.

PROPERTIES OF THE EQUILIBRIUM WITH LARGE NUMBERS OF MARKET MAKERS

In equilibrium, both market makers and informed speculators earn positive expected profits. Market makers earn positive profits due to their oligopolistic access to the order flow, i.e., their ability to trade on the basis of current prices. Informed speculators earn profits due to their monopolistic access to information. It is clear from equation (17) that $E(\Delta\tilde{p}_2|\tilde{d} + \tilde{g}) = 0$. Thus, it is impossible for a trader who is not a market maker and who has no private information to construct a consistently profitable trading strategy based on public information alone.[6] At the same time, however, the fact that market makers make positive profits implies that prices do not follow a Martingale, i.e., it is not generally true that $E(\tilde{p}_2|\tilde{p}_1) = \tilde{p}_1$. In fact, prices overreact in period one, rising too much (for a Martingale) when the order flow is positive and falling too much when the order flow is negative thus allowing market makers, who face the order flow, to liquidate positions later at a profit on average. As the number of market makers increases, the market power of each market maker is reduced toward zero, the profits of each market maker (indeed, of market makers as a whole) tend to vanish, and the Martingale property holds in the limit as $M \to \infty$.

In the rest of this paper we assume that M is approximately infinite, so that the Martingale property holds.

Defining λ^* as the limit of λ as $M \to \infty$, we have:

$$\lambda^* = \frac{N^{1/2}(\frac{1}{\tau_0+\tau_G} + \frac{1}{\tau_1})^{1/2}}{2(N+1+\frac{2\tau_I}{\tau_0+\tau_G})(\sigma_u^2)^{1/2}} \text{ , and} \tag{26}$$

$$\beta = \frac{(\sigma_u^2)^{1/2}}{N^{1/2}(\frac{1}{\tau_0+\tau_G} + \frac{1}{\tau_I})^{1/2}} . \tag{27}$$

The equilibrium price in period one is given explicitly by:

$$\tilde{p}_1 = \tilde{p}_0^* + \lambda^*(\beta \sum_{n=1}^{N} \tilde{i}_n + \tilde{u}) \tag{28}$$

$$= p_0^* + \frac{1}{2(N+1+(\frac{2\tau_I}{\tau_0+\tau_G}))} \cdot \sum_{n=1}^{N} \tilde{i}_n$$

$$+ \frac{N^{1/2}(\frac{1}{\tau_0+\tau_G} + \frac{1}{\tau_I})^{1/2}}{2(N+1+\frac{2\tau_I}{\tau_0+\tau_G})} \cdot \frac{\tilde{u}}{(\sigma_u^2)^{1/2}} . \tag{29}$$

How informative are prices in this equilibrium? Define $\tau(p_1)$, the precision of prices by $\tau(p_1) = var^{-1}(\tilde{d}|\tilde{p}_1)$. It is a tedious but straightforward exercise to show that:

$$\tau(p_1) = \tau_0 + \tau_G + \left(\frac{\tau_0+\tau_G}{2(\tau_0+\tau_G)+\tau_I}\right)N\tau_I . \tag{30}$$

This formula for $\tau(p_1)$ is quite useful. For the sake of comparison, define τ^* as the precision of the best estimate of $\tilde{d}$ which could be made on the basis of all public and private information. It is easy to show that:

$$\tau^* = \text{var}^{-1}(\tilde{d} \mid \tilde{d} + \tilde{g}, \tilde{d} + \tilde{e}_1, \ldots, \tilde{d} + \tilde{e}_N) \qquad (31)$$

$$= \tau_0 + \tau_G + N\tau_I.$$

By comparing the formulas for $\tau(p_1)$ and τ^* in the two above equations, we see that the price $\tilde{p}_1$ does not reflect all socially available information; intuitively, we can say that to the prior precision τ_0, the price adds all of the public precision τ_G but only a fraction of the private precision $N\tau_I$. This fraction which is less than 1/2, is given by $(\tau_0 + \tau_G)/(2\tau_0 + 2\tau_G + \tau_I)$.

Clearly, it is by withholding some of their private information from the market that informed speculators are able to make profits on average. It can be shown that the expected profits of each informed trader, which we denote π, are given by:

$$\pi = \frac{(\tau_0 + \tau_G + \tau_I)\tau_I \sigma_u^2}{4N[2(\tau_0 + \tau_G) + (N+1)\tau_I](\tau_0 + \tau_G)} \qquad (32)$$

We can make the number of informed speculators endogenous by assuming that each informed speculator must pay a cost c to acquire this private information before random variables are realized. Then the number of informed speculators will be approximately the number N which makes $\pi = c$. In this equilibrium, the losses of noise traders, which are equal to the trading profits of informed speculators, are translated dollar for dollar into resources spent acquiring private information, no more than half of which is incorporated into prices.

THE EFFECT OF PUBLIC INFORMATION ON THE INFORMATIVENESS OF PRICES

How does a change in the amount of public information affect the informativeness of prices? In our model, answering this question is a comparative statistics exercise concerning the effect of increasing τ_G.

If the number of speculators is held constant, the answer to the question can be obtained by inspecting equation (30). Clearly, an increase in τ_G not only increases prices directly via the announcement effect but also indirectly by increasing the percentage of informed traders information which is incorporated into prices. Additional public information tends to increase the efficiency with which private information is incorporated into prices.

When the number of speculators is endogenous, we can see from equation (32) that an increase in τ_G tends to reduce π and thus leads to an exit of speculators.

We wish to know whether the increase in τ_G can reduce the number of speculators so much that the informativeness of prices is reduced as a result of providing more information publicly. The expression for $\tau(p_1)$ in equation (30) can be written:

$$\tau(p_1) = \frac{2(\tau_0 + \tau_G) + (N+1)\tau_I}{2(\tau_0 + \tau_G) + \tau_I}\,\tau \qquad (33)$$

Combining this expression with the equilibrium condition $\pi = c$ and equation (32), we obtain:

$$\tau(p_1) = \frac{1}{4}\left[\frac{(\tau_0 + \tau_G + \tau_I)\tau_G \tau_I \sigma_u^2}{[2(\tau_0 + \tau_G) + \tau_I]^2\,N\,c^2}\right]. \qquad (34)$$

Since an increase in τ_G reduces N and since the partial derivative of the right-hand-side of the above equation with respect to τ_G is positive, while the partial derivative with respect to N is negative, it is clear that an increase in τ_G increases the informativeness of prices $\tau(p_1)$, even though it reduces the amount of information produced privately.

This result provides a justification for the government to produce information and to release it publicly to all traders simultaneously. If the government produces information at the same cost as private speculators, there are clear advantages to having public production of information. Even if the government is a higher cost producer of information than private speculators, there may be advantages to having the government produce private information because of the better efficiency with which it is incorporated into prices and its effect on the incentives of private speculators.

THE EFFECT OF UNINFORMED NOISE TRADING ON THE INFORMATIVENESS OF PRICES

An increase in uninformed noise trading can be modeled in comparative statics terms as an increase in σ_u^2. With a fixed number of informed speculators, it is clear from equation (30) that an increase in noise trading has no effect on the informativeness of prices at all.

The reason that σ_u^2 does not affect the informativeness of prices is that market makers and informed speculators scale up their activities proportionately as σ_u^2 increases. That is, if the standard deviation σ_u doubles, then each informed speculator doubles the quantity he trades in the new equilibrium. He does this because the quantity he trades is not limited by risk aversion but rather by the quantity the market will bear. When σ_u doubles, the market will bear twice as much because $\lambda*$ is halved. This occurs because market makers also double the quantities they trade. Note that the role of noise trading and liquidity here is completely different from the competitive model of Grossman and Stiglitz (1981), where increases in noise trading and risk aversion interact to make prices less informative.

With endogenous speculators, it can be seen from equation (32) that an increase in uninformed noise trading tends to increase the profits of informed speculators and thus induce entry of new speculators. From equation (30), it can be seen that this increase in N leads to an increase in the informativeness of prices. From the result in Section 3, we conclude that increases in noise trading tend to stabilize prices overall by shifting less volatility into the present than is shifted out of the future and increases the efficiency with which resources are allocated to consumption. The increased efficiency with which resources are allocated to consumption is bought at a cost. That is, society as a whole spends more resources acquiring information than before. These real resources are provided through the trading losses of uninformed noise traders. Within the framework of this model, it thus appears that the consumers of the commodity have no incentive to discourage unprofitable speculation (except perhaps on paternalistic grounds) and may even have an incentive to encourage it.[7]

CONCLUSION

In this paper we have examined a three period model of a speculative market which focuses on the informational effects of speculation. We have shown that in a model where the ability to speculate successfully is based on the possession of private information, and the willingness to hold risky positions does not depend on risk aversion but rather on the ability of the market to supply positions, then an aggregate economy-of-scale property emerges in which uninformed speculation has a stabilizing rather than a destabilizing effect. We have also shown that when information is costly, public generation of information tends to add informativeness to prices. A key feature of the model is the idea that uninformed speculation and hedging to the extent that both are random, should be seen as having the same kind of effect on prices. Furthermore, the uninformed participants who lose money consistently pay for the information which is privately produced and incorporated into prices through speculation. If a willingness on the part of some participants to make expected losses is not present, then there is no reason for the market to exist.

There are a large number of questions about speculative markets which this paper does not address. Examples are the following:

1. Why are hedgers and speculators willing to lose money consistently?
2. What is the optimal size of a speculative firm?
3. What properties does an over-the-counter market without market makers have?
4. What happens when some speculators acquire information about what uninformed traders are doing rather than about the "fundamentals" of demand and supply?
5. Do commodities exchanges maximize seat values by limiting the number of market makers?
6. What happens when the participation of uninformed traders in the market is influenced by the "cost" of trading.
7. Can the model be made dynamic by having non-degenerate trading take place in more than one period?

It seems likely that questions such as these can be effectively discussed within the framework of models similar to the one developed in this paper.

NOTES

1. The alternative view, that the entry barrier raises returns by restricting the flow of floor trading resources onto the floors of commodity exchanges, is not considered here.
2. This resource constraint rules out storage of the commodity between seasons. It also rules out attempts to manipulate prices by destroying stocks of the commodity.
3. See Kyle (1981a) for proofs.
4. For a model in which informed speculators have price contingent trading, see Kyle (1982).
5. See Kyle (1982) for details.
6. A role for such traders does occur if we assume that $E(\tilde{u}) \neq 0$. Then if there is a competitive fringe of traders who are not market makers and have no private information, it can be shown that this fringe will face the predictable component of the noise trading by trading the quantity $-E(\tilde{u})$ and make zero profits in equilibrium. Aggregating this fringe together with the noise traders gives us the zero mean noise term of our model.
7. This model provides another approach to a

problem raised by Friedman (1969).

REFERENCES

Fama, Eugene F. "Efficient Capital Markets: A Review
     of Theory and Empirical Work." Journal of Finance,
     25 (1979): 383-417.
Friedman, Milton. "In Defense of Destabilizing
     Speculation." In The Optimum Quantity of Money
     and Other Essays. Chicago, Aldine, 1969.
Grossman, Sanford J. and Stiglitz, Joseph E. "On the
     Impossibility of Informationally Efficient
     Markets." American Economic Review 70 (1980: 393-
     408.
Kyle, Albert S. An Equilibrium Model of Speculation
     and Hedging. Unpublished Ph.D. Dissertation,
     University of Chicago, 1981a.
     _____. The Efficient Markets Hypothesis and
     the Supply of Speculative Services. Unpublished
     Manuscript, 1981b.
     _____. Equilibrium in a Speculative Market
     with Strategic Informed Trading. Unpublished
     Manuscript, 1982.
Muth, John F. "Rational Expectations and the Theory
     of Price Movements." Econometrica 29 (1961): 315-
     335.
Niederhoffer, Victor and M. F. M. Osborne. "Market
     Making and Reversal on The Stock Exchange."
     Journal of the American Statistical Association
     61 (1966): 897-916.
Telser, Lester. "Why There are Organized Futures
     Markets." The Journal of Law and Economics
     24 (1981): 1-22.
Working, Holbrook. "Futures Trading and Hedging."
     American Economic Review 43 (1953): 314-343.
     _____. "Price Effects of Futures Trading."
     Food Research Institute, Vol. 1, No. 1, February
     1960.
     _____. "Tests of a Theory Concerning Floor
     Trading on Commodity Exchanges." Food Research
     Institute Studies, Vol. VII, 1967, Supplement:
     Proceedings of a Symposium on Price Effects of
     Speculation in Organized Commodity Markets.

3
Speculative Storage, Futures Markets, and the Stability of Agricultural Prices

Alexander H. Sarris

INTRODUCTION

Most analyses of the welfare effects of commodity market stabilization focus on the benefits to producers and consumers of eliminating price fluctuations. Early literature on the subject that started with the pathbreaking articles of Waugh (1944) and Oi (1961) and was followed by several articles--notable among which are those of Massell (1969), Hueth and Schmitz (1972), and Turnovsky (1974)--used very simple linear theoretical models of price fluctuations with additive disturbances to show that elimination of fluctuations is generally beneficial to society but the distribution of benefits is a function of the slopes of the demand and supply curves and the sources of the fluctuations.

Recent literature has relaxed several of the early assumptions by incorporating nonlinear supply and demand curves, multiplicative disturbances, alternative expectation assumptions and different models of producer and consumer behavior,[1] but the overall conclusion that elimination of price fluctuations by a costless public buffer stock is beneficial to society has stayed remarkably intact. The recent article by Newberry and Stiglitz (1979) seems to be the only place where this basic conclusion has been questioned. The real contribution of these recent writings has been to provide increasingly sophisticated models for the allocation of the total benefits from stabilization to different groups.

Prominently absent from all this literature are the explicit incorporation of a public buffer stock rule with its expected costs, the consideration of private stockholding behavior, and the incorporation of futures markets. Turnovsky (1978b) considered explicit stabilization rules to show that they indeed stabilized

Alexander H. Sarris, University of Athens, Greece.

the market and increased welfare, but his model does not include private stockholding or futures markets and does not consider the cost of stockholding. In a more recent paper, Turnovsky (1979) attempted to incorporate futures markets in a model of stabilization; but his model of a futures market is quite simple, neglecting the interaction between hedgers, speculators, and private stockholders--which is at the heart of the futures markets literature--and, furthermore, still neglecting the cost of stockholding. Other analyses of private stockholding (e. g., Helmberger and Weaver (1977) and Gardner (1979)) have assumed that private traders are perfect arbitragers which is, however, true only under the assumptions of risk neutrality and infinite wealth. The degree to which the private storage trade approaches perfect arbitrage will be seen to have important implications for market price stabilization.

The purposes of this paper are twofold. First, a model incorporating both a private stockholding activity and an explicit public stabilization rule is analyzed with the purpose of illuminating the interaction of these simultaneous stockholding activities. Second, an explicit model of a futures market consistent with the futures market literature is incorporated in the previous model with the purpose of investigating the effectiveness of public intervention in the presence of the futures market.

The results of the analysis are the following. On the one hand, it is shown that, given some private stockholding behavior, the government stabilization policy can indeed reduce the market price variance but at ever increasing net cost as larger degrees of stabilization are sought. As the willingness, however, of the private storage trade to arbitrage expected price differences is increased, the effectiveness of public policy is lessened.

The introduction of futures markets leads to some rather startling results. It is shown that the futures market reduces the variance of spot prices. As far as the author is aware, this result has not been shown analytically before, although Peck (1976), using a different model of the futures market than the one utilized here, showed it for a limited range of some parameters. It is also shown that in the presence of futures markets the role of public intervention becomes smaller.

If the producers use the futures price rather than their (presumably rational) expectations to plan production, then this has additional stabilizing influence on the spot price. However, it seems that stabilization of the futures price by the government is a less desirable way to stabilize the spot markets than direct intervention.

While the major part of the analysis is carried out using a model of a storable commodity that is produced with a lag (fitting most agricultural products), the results carry over to storable commodities that can be produced instantaneously (such as metals and minerals); and in a final section it is shown how the extension can be made.

A MODEL FOR A STORABLE COMMODITY MARKET WITHOUT FUTURES

The model utilized throughout the paper is similar to the one originally used by Muth (1961) and later utilized by Turnovsky (1978b, 1979) and Peck (1976).

A commodity market model is specified that is composed of demand, D_t, supply, S_t, that takes one period to materialize, private speculative storage relations, I_t, and a market clearing identity.

$$D_t = -a\, p_t + u_t \qquad\qquad a \geq 0 \qquad\qquad (1)$$

$$S_t = b\, p^*_{t,t-1} + v_t \qquad\qquad b \geq 0 \qquad\qquad (2)$$

$$I_t = \alpha\, (p^*_{t+1,t} - p_t) \qquad\qquad \alpha > 0 \qquad\qquad (3)$$

$$D_t + I_t = S_t + I_{t-1}. \qquad\qquad\qquad\qquad (4)$$

The model above is specified in deviation form. All variables are differences of the actual magnitudes from those which would prevail if u_t, v_t, the random variables, are set to zero. The expected spot price, in period t as viewed from period t - 1 is $p^*_{t,t-1}$. Stochastic disturbances, u_t and v_t are independently distributed through time with the properties:

$$E(u_t) = 0, \qquad\qquad E(v_t) = 0 \qquad\qquad (5)$$

$$E(u_t^2) = \sigma_u^2, \qquad E(v_t^2) = \sigma_v^2, \qquad E(u_t v_t) = 0 \quad (6)$$

Equations (1) and (2) are straightforward. In equation (3), I_t, is the difference between what is carried over into the next production period and the normal working carry-overs. The linear relation in (3) was derived by Muth by analyzing speculators' optimizing behavior under some approximations. It can also be considered as a linear approximation (for reasonably small variations) about normal working carry-overs to the supply of storage curve that has been analyzed by Working (1949) and Brennan (1958). Figure 1 illustrates the point that different assumptions about what constitutes normal working carry-overs give rise to different values of α.[2/] In the figure the horizontal axis represents $\bar{I} + I_t$,

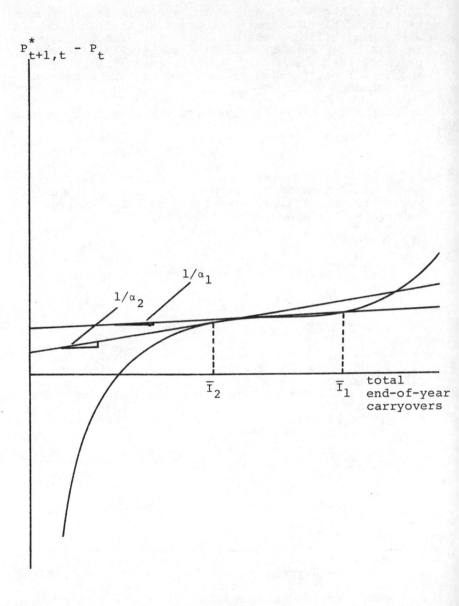

Figure 1   Linear Approximations to the Supply of Storage
Curve Under Different Assumptions of
Working Inventories

namely, the total (working plus speculative) carry-overs, and the straight lines represent the linearizations about $\bar{I}$. Perfect arbitrage would imply that $\alpha = \infty$ in this model so as to assure $p^*_{t+1,t} - p_t = 0$ with finite speculative carry-overs, I_t. Examining Figure 1, it can be seen that $\alpha = \infty$ might be a realistic assumption if $\bar{I}$, the normal level of stocks, is large (e. g., $\bar{I}_1$ in Figure 1). If the normal carry-overs, however, are close to the steeply declining portion of the curve near the vertical axis (e. g., $\bar{I}_2$), it is clear that α should be considered as finite. The $\alpha \approx \infty$ assumption might be a reasonable one for intrayear analysis when uncertainties about production are absent and the level of stocks that have to be carried into the next (within the crop year) period are reasonably well known and large. At the end of the crop year, however, the level of working stocks are low; and normally $\bar{I}$ in Figure 1 will be close to the steeply declining portion. This situation was considered "normal" in the writings of Working (1934, 1948).[3]/ Since in this paper the concern is with interyear price fluctuations, the assumption $\alpha < \infty$ appears reasonable.

Substituting (1)-(3) in the equilibrium relation (4) and rearranging terms, we obtain:

$$-\alpha p^*_{t+1,t} + (\alpha + b)\, p^*_{t,t-1} + (\alpha + a)\, p_t - \alpha\, p_{t-1}$$

$$= u_t - v_t \equiv e_t \qquad (7)$$

where the last equality is a mere convenient definition.

The agents will be assumed to hold rational expectations. Interestingly enough, many of the major results of the paper hold under adaptive expectations as well and this will be pointed out throughout the exposition.

Taking conditional expectations of (7), given information available at time $t - 1$, we obtain:

$$-\alpha\, p^*_{t+1,t-1} + (\alpha + b)\, p^*_{t,t-1} + (\alpha + a)\, p^*_{t,t-1}$$

$$- \alpha\, p_{t-1} = 0. \qquad (8)$$

The solution to (8), as pointed out by Muth and Turnovsky (1979), is given by:

$$p^*_{t+k,t} = r^k \cdot p_t \qquad (9)$$

where r is the smallest root, which is positive and between 0 and 1, of the quadratic equation:

$$-\alpha\, r^2 + (a + b + 2\alpha)\, r - \alpha = 0. \qquad (10)$$

Substituting (9) in (7) and solving for p_t, we obtain:

$$p_t = r \ p_{t-1} + \frac{e_t}{[a + \alpha \ (1 - r)]} \tag{11}$$

It can readily be established (see also Turnovsky (1979)) that the asymptotic variance of p_t is given by the expression:

$$\sigma_p^2 = \frac{\sigma_e^2}{(1 - r)^2 [a + \alpha (1 - r)]^2} \tag{12}$$

where $\sigma_e^2 \equiv \sigma_u^2 + \sigma_v^2$.

The following properties of σ_p^2, which are important for parts of the subsequent analysis, are shown in Appendix A.

$$\frac{\partial \sigma_p^2}{\partial a} < 0, \qquad \frac{\partial \sigma_p^2}{\partial \alpha} < 0 \tag{13}$$

These properties of σ_p^2 also hold under adaptive expectations (see Turnovsky (1979)). What they signify is that (i) the "flatter" the market demand curve the smaller will be the interyear price fluctuations and (ii) the more responsive are private stockholders to interyear arbitrage opportunities, a property that will hold the less risk averse and the wealthier private speculators become, the smaller will be the year-to-year price fluctuations. Both of these are intuitively reasonable properties. Two other interesting properties of σ_p^2 are that

$$\lim_{a \to \infty} \sigma_p^2 = 0, \qquad \lim_{\alpha \to \infty} \sigma_p^2 = 0 \tag{14}$$

To see the first one of these, notice that as $a \to \infty$, $r \to 0$, and the property becomes apparent from (12). The proof of the second one is not very apparent because as $\alpha \to \infty$, $r \to 1$, and is relegated to Appendix B.

The first one of the properties in (14) makes good economic sense because it says that if the market demand curve is infinitely elastic, then quantity fluctuations will not affect the price--an obvious consequence.

The economics of the second property are a little surprising. If $\alpha \to \infty$, then there is perfect interyear arbitrage. This and the property of rational expectations assure that the interyear spot price is perfectly stable. Interestingly, under adaptive expectations it can be shown that $\lim_{\alpha \to \infty} \sigma_p^2 < \infty$. The empirical evidence suggests that $\sigma_p^2 < \infty$. Should this be construed as a refutation of the rational expectations hypothesis? I think not, especially in light of the earlier discussion of this section about the finiteness of α. In other words, the more unrealistic assumption would be that $\alpha \to \infty$ and not necessarily the rational expectations hypothesis.

THE EFFECTS OF PUBLIC PRICE STABILIZATION

Suppose now that for some reason the commodity market analyzed in the previous section is deemed as exhibiting unduly large price fluctuations and action is taken by the government to lessen the fluctuations. The benefits to consumers and producers from doing so have been analyzed extensively in earlier literature as was pointed out in the introduction. The emphasis here will be on the interaction between the public and the private storage.

The form of the rule that is followed by the public authority is assumed to be very simple. Let y_t be the net accumulation or decumulation of public stocks in year t. The rule that will govern these yearly actions is assumed to be given by the following expression:

$$y_t = -\mu p_t, \qquad \mu > 0. \qquad (15)$$

The form of the rule (15) is such as to be naturally stabilizing. The authority accumulates stocks when the market price is below the underlying certain average price and sells them when the market price is above it.

There are two assumptions that are used to justify the use of such a rule. The first one is that when $p_t > 0$ there are always enough previously accumulated stocks to sell. This, of course, in reality is not true. However, the net effect of dropping the assumption and using a nonlinear rule of the form, e. g., $y_t = \max(-\mu p_t, -I_{t-1}^g)$ where I_{t-1}^g is the accumulated public stock at the end of the previous period, will be to increase slightly the resulting asymptotic price variance and to skew the asymptotic price distribution slightly to the right. While these might be interesting empirical effects that can easily be accounted for in empirical investigations of buffer-stock policies,[4] they will not alter the general theoretical results of this paper. Hence, the simpler and much more manageable theoretical rule (15) is adopted.

The second assumption is more fundamental. A rule such as (15) assumes that the government knows the underlying certainty price in year t. If the commodity market is stationary in the sense that the underlying supply demand and storage parameters, the certainty price, and the variances of the random quantity stocks σ_u^2 and σ_v^2 are not changing over time, then this assumption is reasonable since, given enough time, these underlying magnitudes could be estimated. If, however, the commodity market undergoes frequent structural shifts, then it is quite likely that the government will err in its estimation of the underlying structure and these errors might be large. One possible source of errors is that the attempt to stabilize the price might have an influence on the average quantity produced and lead to a secular shift in the underlying average price which is very hard to estimate. Turnovsky (1978b), in fact, has shown that the government rule might destabilize the spot price if the variance of the errors is large enough.

Given these caveats, the rule in (15) is an idealization of what the public sector could do in stabilizing the market. Nevertheless, it will be shown that even this ideal rule has its downbacks.

Given (15), the only equation that changes in the original model is the market clearing identity which becomes:

$$D_t + I_{t-1} + y_t = S_t + I_t. \tag{16}$$

Substituting the appropriate expressions and rearranging terms, the following equation is obtained:

$$-\alpha \ p^*_{t+1,t} + (\alpha + b) \ p^*_{t,t-1} + (\alpha + a + \mu)p_t$$

$$- \alpha \ p_{t-1} = e_t. \tag{17}$$

Comparing this with (7) it can be seen that the effect of the stabilizing public rule is to increase the size of the parameter a. In other words, the public rule "rotates" the market demand curve about the underlying certainty price and makes it flatter. Given the result of the previous section that $\partial\sigma^2_p/\partial a < 0$, it is clear that the introduction of the rule has a stabilizing tendency on the price variance as expected. The result of the previous section that $\lim_{a\to\infty} \sigma^2_p = 0$ interpreted in the current context means that the government can stabilize the spot price perfectly, but this necessitates infinite μ, in other words, infinite capacity.

Suppose now that the government wants to set μ so as to reduce the asymptotic spot price variance to some finite level, say, σ^{*2}_p. The monotonicity of σ^2_p (μ), which is evidenced by the property:

$$\frac{d\sigma^2_p}{d\mu} = \frac{\partial\sigma^2_p}{\partial a} \frac{da}{d\mu} = \frac{\partial\sigma^2_p}{\partial a} < 0,$$

ensures that there is indeed a unique such $\mu^*(\alpha)$.

The following property of μ^* is proven in Appendix C.

$$\frac{\partial\mu^*(\alpha)}{\partial\alpha} < 0 \tag{18}$$

The economics of (18) are quite interesting. They say that as the private storage trade is <u>more</u> responsive to interyear arbitrage opportunities, the <u>less</u> government intervention is needed to achieve a given degree of price stabilization (or, equivalently, a given reduction in spot price variance).

Equation (18) might not hold under adaptive expectations. To see this, assume that expectations are formed according to the equation:

$$p^*_{t+1,t} = \beta \cdot p_t + (1 - \beta) \, p_{t,t-1} \qquad 0 \le \beta \le 1. \quad (19)$$

As Turnovsky (1979) has shown, the introduction of (19) in equation (17) leads to the following expression for the asymptotic spot price variance:

$$\sigma_p^2 = \frac{\{\beta[a + \mu + 2\alpha \, (1 - \beta)] + 2 \, (a + \mu + b)}{(a + \mu) \, (a + \mu + b) \, [(a + \mu) \, (2 - \beta)}$$
$$\frac{(1 - \beta)\} \, \sigma_e^2}{- \beta b + 4\alpha \, (1 - \beta)]}. \quad (20)$$

As $\alpha \to \infty$ in the above expression, it can be seen that:

$$\lim_{\alpha \to \infty} \sigma_p^2 = \frac{\gamma \, \sigma_e^2}{2 \, (a + \mu) \, (a + \mu + b)}. \quad (21)$$

The limiting price variance, as the private storage trade tends toward perfect arbitrage, is not zero unless $\beta = 0$. Hence, it follows that, even if $\alpha \to \infty$, there is room for further reduction in the price variance by government intervention.

If the government wants to reduce the variance to σ_p^{*2}, it must set μ at a level $\mu^0(\sigma_p^{*2}/\sigma_e^2, \alpha, \gamma)$. By differentiating (20) and rearranging terms, it can be shown that:

$$\frac{\partial \mu^0}{\partial \alpha} = \frac{2 \, (1 - \beta) \, [\beta - 2 \, k}{(2 - \beta) \, [k \, (2a + 2\mu^0 + b) + k \, (a + \mu^0)}$$

$$\frac{(a + \mu^0)}{(a + \mu^0 + b) - 1] + k \, (2a + 2\mu^0 + b)}$$

$$\frac{(a + \mu^0 + b)]}{[4\alpha \, (1 - \beta) - \beta b]}$$

where

$$k \equiv \frac{\sigma_p^{*2}}{\sigma_e^2}. \quad (22)$$

It can be seen from (22) that, depending on the value of β and k, $\partial\mu^0/\partial\alpha$ can be positive or negative; hence, (18) does not hold universally.

The above analysis shows that the form of expectations is important in determining whether more or less government intervention is needed to afford a given degree of price stability when the private trade becomes more responsive to arbitrage opportunities.

THE NET COST OF PRICE STABILIZATION

All previous theoretical literature has assumed that the cost to the government of stabilizing the spot price is zero or, in other words, that the storage cost is made up by the income that the public authority makes by buying low and selling high. It is shown in this section that this is generally not so.

First, consider the income of the stocking agency. In year t the agency receives (spends) an amount R_t equal to:

$$R_t = -y_t\,(p_t + P) \tag{23}$$

where P is the underlying certain price. Substituting for y_t from (15), we obtain:

$$R_t = \mu\,p_t^2 + \mu\,P\,p_t. \tag{24}$$

The average revenue of the public agency per unit of time is hence given by:

$$R(\mu) = \frac{\mu\,\sigma_e^2}{(1 - r_\mu^2)\,[a + \mu + \alpha\,(1 - r_\mu)]^2}. \tag{25}$$

It is immediately obvious that $R(0) = 0$. As $\mu \to \infty$, the numerator as well as the denominator tend to infinity (see Appendix B), and the magnitude of R is not obvious.

In Appendix D it is shown that:

$$\lim_{\mu\to\infty} R(\mu) = 0. \tag{26}$$

Since $R(\mu)$ is finite for all other values of μ, the shape of $R(\mu)$ must be as shown in Figure 2 by curve R.

To analyze the average storage cost to the public authority per unit of time, $S(\mu)$, suppose that intervention starts in some period which we denote as period 1. It is necessary, of course, that $p_1 < 0$. The public stocks at the end of period 1 are y_1, and at the end of period t it can be seen that they are equal to $\Sigma_{i=1}^{t}\,y_i$. If it is assumed that the cost of storing a unit of the commodity for one period is constant and equal to c, then the cost to the government of carrying the program for t periods is equal to:

$$c_t = c \left[y_1 + (y_1 + y_2) + \ldots + \sum_{i=1}^{t} y_i \right]$$

$$= c \left[t y_1 + (t - 1) y_2 + \ldots + 2 y_{t-1} + y_t \right] \qquad (27)$$

$$= -c \mu \left[t p_1 + (t - 1) p_2 + \ldots + 2 p_{t-1} + p_t \right]$$

From the price equation (11) it can be seen that:

$$p_t = r_\mu^{t-1} p_1 + \frac{1}{\ell} \sum_{i=0}^{t-2} r_\mu^i e_{t-i} \qquad (28)$$

where $\ell \equiv [a + \mu + \alpha (1 - r_\mu)]$, and the symbol r_μ denotes the fact that r is a function of μ.

Substituting (28) in (27) and collecting terms, we obtain:

$$c_t = -c\mu \left[p_1 \sum_{j=1}^{t} j \, r_\mu^{t-j} + \frac{1}{\ell} \sum_{j=1}^{t-1} j \sum_{i=0}^{t-1} r_\mu^i e_{t-j-i+1} \right] \qquad (29)$$

Consider now many repetitions of all possible storage programs that last exactly t periods. It is clear that on average the second term in the brackets of (29), that includes the e_i's which are not restricted as to sign, will average to zero. On the other hand, p_1 must always be negative; otherwise a storage program cannot start and, hence, on the average p_1 will be equal to the mean of the conditional density of p_t conditioned by $p_t < 0$. Denote this conditional mean by $-p$ ($p > 0$). The above reasoning then leads to the conclusion that the storage cost of programs that last t periods on average is equal to:

$$\overline{c}_t = c\mu p \sum_{j=1}^{t} j \, r_\mu^{t-j}. \qquad (30)$$

The summation terms in (30) can be shown to be equal to the following:

$$\sum_{j=1}^{t} j \, r_\mu^{t-j} = \frac{(1 - r_\mu)(1 + t) - (1 - r_\mu^{t+1})}{(1 - r_\mu)^2} \qquad (31)$$

Consider now our objective, the cost per unit of time and its limit as $t \to \infty$.

$$C(\mu) \equiv \lim_{t \to \infty} \frac{\overline{C}_t}{t} = \lim_{t \to \infty} \; c\mu p \left[\frac{1 + t}{(1 - r_\mu) t} - \right.$$

$$\left. - \frac{1 - r_\mu^{t+1}}{(1 - r_\mu)^2 \, t} \right] = \frac{c\mu p}{1 - r_\mu}. \qquad (32)$$

From (32) it is easy to show that:

$$C(0) = 0, \qquad C(\infty) = \infty, \qquad \frac{\partial C(\mu)}{\partial \mu} > 0. \qquad (33)$$

Hence, the average yearly cost of carrying out a price stabilization program increases ad infinitum with the increased government intervention. This is shown graphically in Figure 2 by curve C.

It is clear that price stabilization can become a very costly proposition even under the idealized--and favorable to the government--conditions that are assumed here. Since the profits to the public agency are always bounded while the costs become increasingly large for larger degrees of price stabilization, it is clear that the net costs can increase indefinitely. Hence, large degrees of price stabilization, while always concurring finite social benefits as has been shown in earlier literature, have increasingly large social costs. Therefore, perfect price stabilization is never socially optimal.

FUTURES MARKETS AND PRICE STABILIZATION

Suppose now that the previously analyzed commodity market is supplemented by an organized futures market. There are two important elements that are introduced in such a case. First, the possibility of hedging by holders of physical quantities arises. Second, the futures market attracts speculators who do not deal in the cash market.

The simultaneous determination of spot and futures prices has been treated by Stein (1961) in a one-period model, and we use a linearized multiple period version of his model with one additional restriction.

As Stein has pointed out, for unchanged expectations the introduction of the futures market leads to an increase in total physical stocks held because now there are some stocks that are hedged, while before all the stocks were unhedged. Denote by $p^f_{t+1,t}$ the price at time t of a futures contract due to expire at time $t + 1$. Then the net quantity of hedged inventories is assumed to be:

$$H_t = h \, (p^f_{t+1,t} - p_t) \qquad h \geq 0 \qquad (34)$$

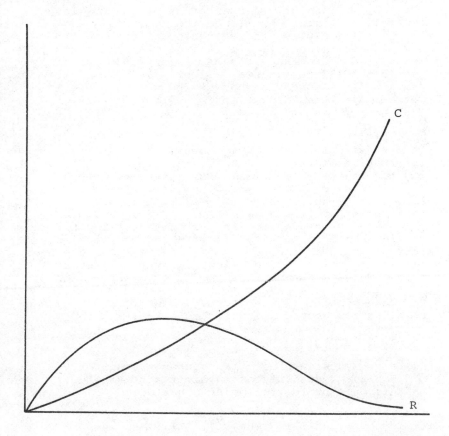

average
cost per
unit time

Figure 2   Shapes of Expected Revenue and Cost Curves
per Unit Time of the Public Stabilization
Authority

Equation (34) makes net hedging a function of the current basis (the difference between current futures and spot price). It is a linearized version of Stein's hedging equation which in our notation is:

$$H_t^s = H\ [(p_{t+1,t}^* - p_{t+1,t}^{f*}) + (p_{t+1,t}^f - p_t) - c]$$

$$H' > 0 \qquad (35)$$

where $p_{t+1,t}^{f*}$ is the expected price of the futures contract at time $t + 1$ viewed from time t.

In the analysis here we neglect storage costs (which under linearity will cancel out in the analysis anyway), and we make the reasonable assumption that in period $t + 1$ (the futures delivery period) the futures price will be equal to that period's cash price. Notice that this assumption is _not_ the same as:

$$p_{t+1,t}^f = p_{t+1,t}^* \qquad (36)$$

In our assumption $p_{t+1,t}^*$ is equated to $E\ (p_{t+1,t}^f)$ given information at time t. The distinction is subtle, but its importance is crucial. For instance, Turnovsky (1979) assumed an equation like (36), making his analysis one not of futures markets but instead of rational expectations.

While our assumption is noncontroversial, the truth of (36) has been the subject of a rather voluminous interature under the rubric of bias of futures prices.[5]/ The model analyzed here allows for a risk premium. All we assume here is that, as the futures expiration date comes close, this risk premium must tend to zero.

The speculators in futures are assumed to buy and sell futures according to the relation:

$$F_t = f\ (p_{t+1,t}^* - p_{t+1,t}^f) \qquad f > 0 \qquad (37)$$

where $F_t > 0$ signifies a long futures position.[6]/

The total quantity of physical stocks held is given in this case by the following equation:

$$I_t = \alpha\ (p_{t+1,t}^* - p_t) + h\ (p_{t+1,t}^f - p_t). \qquad (38)$$

The supply, demand, and spot market equilibrium relations remain unchanged.

There are two simultaneous markets that must clear now, given by the following market clearing identities:

$$D_t + I_t = S_t + I_{t-1}. \qquad (39)$$

$$F_t = H_t. \qquad (40)$$

Substituting (37) and (34) in (40) and solving for $p_{t+1,t}^f$, we obtain an expression for the futures price:

$$p^f_{t+1,t} = \frac{h\,p_t + f\,p^*_{t+1,t}}{h + f}\;. \tag{41}$$

Notice that while $E\,(p^f_{t+1,t}) = p^*_{t+1,t}$ holds always, the relation $p^f_{t+1,t} = p^*_{t+1,t}$ will hold only if $h = 0$ or, in other words- if there is no hedging. If there is no hedging, however, there is no physical linkage between the futures and the cash markets, and the futures market is a purely speculative one. In such a case there is no reason why the expected price that is used by producers should be equal to the expected price that is used by the speculators. We have a Keynesian beauty contest problem in the futures market while it might not be true in the actuals market. As will be seen shortly, this has striking consequences for the stability of the spot prices.

Substituting now the expression in (41) into equation (38) and then the expressions for D_t, S_t, I_t, and I_{t-1} into equation (39) after rearranging terms we arrive at the following equation:

$$-\left(\alpha + \frac{hf}{h+f}\right)p^*_{t+1,t} + \left(\alpha + \frac{hf}{h+f} + b\right)p^*_{t,t-1}$$

$$+\left(\alpha + \frac{hf}{h+f} + a\right)p_t - \left(\alpha + \frac{hf}{h+f}\right)p_{t-1} = e_t. \tag{42}$$

Comparing equation (42) with equation (8) for the uninterfered market without futures, it can be seen that the impact of the introduction of the futures markets, as far as the price equation is concerned, is to increase the value of the parameter α to $\alpha + hf/(h + f)$.

As was shown, however, in Section I, such an increase in α will lessen the asymptotic price variance σ^2_p of the interyear spot price. This conclusion, furthermore, is independent of whether we assume rational or adaptive expectations formation.

The above conclusion is strong, and depends crucially on the assumption that h, $f > 0$. If either of these parameters is zero, the spot price variance is unaffected by the introduction of the futures market. In other words, a futures market has stabilizing effects on the interyear spot market if it is used as a hedging medium by the physical handlers of the commodity and if there are some speculators to assume the risks.

Notice that nothing has been said about the intrayear spot price movements. It is in principle possible that a futures market can destabilize the intrayear spot-price variance while stabilizing the year-to-year one, but the investigation of intrayear price variability is beyond the scope of this paper.

Since the quantity $z = hf/h + f$ can be written as

$z = 1/(1/h + 1/f)$, it can be seen that an increase in h or f will increase the value of z and, hence, the value of α leading to the decrease in the asymptotic variance of the spot price. This observation brings out the importance of futures speculators for price stability. If the futures market does not attract enough speculators (f small), then there is not enough opportunity for risk sharing for the holders of physical stocks; hence, they tend to hold less, thus leading to less interyear smoothing of prices.

The impact of government stabilization is quite easy to investigate in this context, given the results of earlier sections. As was pointed out earlier, a public storage rule of the form $y_t = -\mu\, p_t$ only modifies the value of the parameter a to a + μ in equation (41) and leads to a decrease in the value of the asymptotic interyear spot price variance.

As was also pointed out in equation (18), for given desired degree of price stabilization, the amount of government intervention is less the higher is the value of the parameter α if expectations are formed rationally. It was just seen, however, that a properly functioning futures market will increase the value of α. Hence, in the presence of a futures market that is used as a hedging medium and if expectations are rational, there is less public intervention required in order to achieve a given degree of price stabilization than if a futures market were absent. This result, of course, depends on the amount of stabilization sought in the case when expectations are adaptive, as discussed earlier.

THE USE OF FUTURES PRICES BY PRODUCERS

In the previous sections the producers were assumed to base their production plans on expected prices. In the presence of a futures market, however, there is a readily and cheaply available forward price whose use could save the producer the effort of trying to form his own expectations. In this section we investigate the consequences of such use of the futures price on the spot price variance.

Assume that a fraction, $\delta\,(0 \leq \delta \leq 1)$ if the producers base their production plans on the futures price, while the rest form their own expectations as before. Then the model of supply in the previous sections [equation (2)] is modified as follows:

$$S_t = b \left[\delta\, p^f_{t,t-1} + (1 - \delta)\, p^*_{t,t-1} \right] + v_t. \qquad (43)$$

The model of the futures market remains as in the previous section and, hence, equations for inventory carry-overs and futures price are as in (38) and (41).

Substituting in the market-clearing identity (39) and eliminating the futures price via (41), we obtain after some rearrangement the following price equation.

$$-\left(\alpha + \frac{hf}{h+f}\right) p^*_{t+1,t} + \left[\alpha + \frac{hf}{h+f} + b\left(1 - \frac{\delta f}{h+f}\right)\right]$$

$$p^*_{t,t-1} + \left(a + \alpha + \frac{hf}{h+f}\right) p_t$$

$$-\left(\alpha + \frac{hf}{h+f} - \frac{\delta h}{h+f}\right) p_{t-1} = e_t. \tag{44}$$

As expected for $\delta = 0$, (44) reduces to (42). Define:

$$\eta \equiv \alpha + \frac{hf}{h+f} \tag{45}$$

$$\theta \equiv \frac{\delta h}{h+f}. \tag{46}$$

Under rational expectations and following the earlier analysis, it can be shown that:

$$p^*_{t+k,t} = r^k p_t \tag{47}$$

where r is the smallest root of the quadratic equation:

$$\eta r^2 - [a + 2\eta + b(1 - \theta)]r + (\eta - b\theta) = 0 \tag{48}$$

It should be noted that the two roots of (48) are not guaranteed to be both positive unless:

$$\eta - b\theta > 0. \tag{49}$$

If (49) does not hold (which might be likely for large values of δ), then one root is smaller than one in absolute falue and negative (the other root is larger than one); in this case the price will follow a stable cobweb pattern. Increasing values of δ imply increasing naiveté by the producers who prefer to take the market-determined observable forward price more seriously than the unobservable rational expectation; hence, cobweb behavior is not to be ruled out.

A sufficient (but not necessary) condition for (49) to hold, however, is $\alpha > b$, and this might be a reasonable assumption since it implies that speculative inventory holders are more eager to take advantage of arbitrage opportunities than the producer is willing to vary his supply based on expected price-level changes. In any case, all that the following analysis requires is that $|r| < 1$.

Substituting (47) in (44) and following the earlier analysis, it can be shown that the asymptotic yearly variance is given by:

$$\sigma_p^2 = \frac{\sigma_e^2}{(1 - r^2) [a + \eta (1 - r)]^2} . \qquad (49)$$

It is shown in Appendix E that:

$$\frac{\partial \sigma_p^2}{\partial \delta} < 0. \qquad (50)$$

Hence, increasing use of the futures price as a planning price by the farmers has a stabilizing influence on the year-to-year price.

As early as 1947, D. Gale Johnson--in his book, Forward Prices for Agriculture--argued that the establishment of forward prices would reduce some of the instability faced by farmers. The result here essentially supports his original conjecture, if the futures markets are indeed appropriate forward-pricing institutions and if the farmers take them into account for production decisions. Notice that the farmers do not have to be active participants in the market (as hedgers or speculators) for the result to hold.

SHOULD THE GOVERNMENT STABILIZE THE FUTURES PRICES?

It has been advocated by Houthakker (1967) that stabilization of the cash market could be achieved easier if the government, instead of buying and selling physical stocks, bought and sold futures contracts. It is shown here that, under some conditions, the government might actually destabilize the spot price while, under other conditions, it can eliminate part but not all of the spot price fluctuations by intervening in the futures market but at a potentially very large net cost.

Denote by F_{gt} the holdings at time t of futures contracts that mature at t + 1 by the government (F > 0 denotes a long position). The rule that the government will be assumed to follow is of similar form as (15):

$$F_{gt} = -\lambda \ p_{t+1,t}^f. \qquad (51)$$

This rule assumes that the government knows the underlying certain market price and, hence, is an idealization of an actual desirable, realistic rule.

The clearing equation for the futures market becomes:

$$F_t + F_{gt} = H_t, \qquad (52)$$

leading to the following expression for the futures price [compare with (41)],

$$p_{t+1,t}^{f} = \frac{h\ p_t + f\ p_{t+1,t}^{*}}{h + f + \lambda} \ . \tag{53}$$

Substituting (53) in (38) and using (2) for the supply equation, the market-clearing identity (39) yields the following price equation:

$$-\left(\alpha + \frac{hf}{h + f + \lambda}\right) p_{t+1,t}^{*} + \left(\alpha + \frac{hf}{h + f + \lambda} + b\right)$$

$$p_{t,t-1}^{*} + \left(a + \alpha + \frac{hf + h\lambda}{h + f + \lambda}\right) p_t -$$

$$\left(\alpha + \frac{hf + h\lambda}{h + f + \lambda}\right) p_{t-1} = e_t. \tag{54}$$

As expected for $\lambda = 0$, (54) reduces to (42). Under rational expectations, the equation for r corresponding to (10) is:

$$\eta'\ r^2 - [a + b + 2\eta' + \rho]\ r + \eta' + \rho = 0 \tag{55}$$

where

$$\eta' \equiv \alpha + \frac{hf}{h + f + \lambda} \tag{56}$$

$$\rho \equiv \frac{h\lambda}{h + f + \lambda} \ . \tag{57}$$

While both roots of (55) are real and positive, there is a possibility that both are larger than one rendering the system unstable. It can be shown (by comparing the smallest positive root to one) that this can happen if:

$$\rho > a + b. \tag{58}$$

From (57) it can be seen that $\rho(0) = 0$, $\rho(\infty) = h$, and $\rho'(\lambda) > 0$; hence, an upper bound of ρ is h. It can then be seen from (58) that a sufficient (but not necessary) condition for stability of the price equation is:

$$h < a + b. \tag{59}$$

It follows that, if (59) does not hold, then it is possible that the attempt of the government to stabilize the futures price might have destabilizing influence on the spot price.

If (59) holds, then a solution to (55) that is positive and smaller than one exists (denote it by r_λ); then the asymptotic spot price variance can be shown to be qual to:

$$\sigma_{p,\lambda}^{2} = \frac{\sigma_e^2}{(1 - r_\lambda^2)\ [a + \eta'\ (1 - r_\lambda) + \rho]^2} \tag{60}$$

In Appendix F, it is shown that:

$$\frac{\partial \sigma_{p,\lambda}}{\partial \lambda} < 0 \tag{61}$$

so that increasing amounts of government futures market stabilization effort does, indeed, in this case (as Houthakker conjectured) lead to increasing stabilization of the spot markets.

Notice, however, now that:

$$\lim_{\lambda \to \infty} \sigma^2_{p,\lambda} > 0 \tag{62}$$

This can be easily seen since $\lim_{\lambda \to \infty} \eta' = \alpha$ and $\lim_{\lambda \to \infty} \rho = h$ and, hence, r assumes a value between zero and one; by (60), the validity of (62) is apparent.

Inequality (62) highlights the fact that, no matter how hard the government tries to stabilize the futures market, it will never achieve complete stabilization of the spot prices. This is in contrast to what was shown earlier, namely that intervention by the government in the real market can achieve (under the idealized conditions assumed) any degree of price stability. Here it was seen that futures market stabilization can reduce the variance of the spot price only within certain bounds.

The average cost per unit of time of operating the futures market rule (51) can be seen to be equal to $\lambda PC'$ where C' is the interest cost of holding an amount of money equal to λP tied up for one period. It can be seen that, as λ becomes very large, the cost also increases proportionately. The unit time profit on the other hand is equal to:

$$R_{ft} = \lambda \, p^f_{t+1,t} \left(p^f_{t+1,t} - p_{t+1} \right) \tag{63}$$

whose expected value is equal to:

$$R_f = \frac{\lambda \, \sigma^2_{p,\lambda} \, (f \, r_\lambda + h)}{(h + f + \lambda)^2} \, [h \, (1 - r_\lambda) - \lambda \, r_\lambda]. \tag{64}$$

For large values of λ, R_f becomes negative and, hence, the net cost of a futures market stabilization program, even under the ideal conditions considered here, can easily run into the red.

A final caveat about the use of the futures markets as substitutes for the spot markets for stabilization purposes is that, when the farmers base part of their expectations on the futures price, then government intervention might destabilize the spot prices.

To see this, one must combine the models of the

previous and the present sections. Without going through the detailed analysis, it can be shown that the sign of $\partial\sigma^2_{p,\lambda}/\partial\lambda$ is equal to the sign of the expression,

$$b\delta - h\ (1 - r'_\lambda)\ [a + b + \eta'\ (1 - r'_\lambda) + \rho + br'_\lambda]\quad(65)$$

where η' and ρ are as defined in (56) and (57), while r'_λ is the smallest root[8] of the quadratic equation:

$$\eta'\ r - \left(a + b + 2\eta' + \rho - b\ \delta\ \frac{h + \lambda}{h + f + \lambda}\right)\ r$$

$$+\left(\eta' + \rho - \frac{b\delta h}{h + f + \lambda}\right) = 0.\quad(66)$$

If $\delta > 0$, it can be seen from (64) that the government intervention, indeed, might increase the asymptotic variance of the year-to-year spot price.

The arguments of this section can be interpreted to mean that indirect public control of the spot price via the futures market might achieve results contrary to what is desired. Hence. unless other arguments are given to the contrary, the direct control of the spot price via physical stocks is to be preferred as far as stabilization is concerned.

COMMODITY MARKETS WITH INSTANTANEOUS PRODUCTION

In this section, it will be shown that most of the results that were derived earlier hold also when the commodity is characterized by a production relation that responds to current rather than future expected prices. Most mineral and metals primary commodity markets are of this nature.

The only relation that needs to be modified is the equation for S_t. We substitute the following relation in the original model (1)-(4) in the place of (2):

$$S_t = bp_t + v_t.\quad(67)$$

Substituting in the market-clearing identity and collecting terms, the following equation is obtained instead of (7):

$$-\alpha\ p^*_{t+1,t} + \alpha\ p^*_{t,t-1} + (\alpha + a + b)\ p_t$$

$$- \alpha\ p_{t-1} = e_t.\quad(68)$$

It is easy to see, following the analysis of Section 1, that under rational expectations the futures expected prices can be expressed in terms of actual prices as:

$$p^*_{t+1,t} = r^k\ p_t,\quad(69)$$

where r is the smallest positive root of an equation

identical to (10).

Substituting (44) in equation (43) and solving for p_t, we obtain:

$$p_t = r\, p_{t-1} + \frac{e_t}{a + b + \alpha\,(1 - r)},\qquad (70)$$

which has an asymptotic variance equal to:

$$\sigma_p^2 = \frac{\sigma_e^2}{(1 - r^2)\,[a + b + \alpha\,(1 - r)]^2}$$

which, as far as the parameters a and α are concerned, has identical structure as σ_p^2 in equation (12). It can thus be seen that all the earlier obtained results as far as the influence of a, α, μ, and futures markets are concerned--carry over to such commodity markets.

It is not difficult to show that, under adaptive expectations, the above model also preserves all the earlier results.

CONCLUSIONS

The results of this paper have important consequences with respect to public policy vis-à-vis price stabilization.

Since it was shown that, in the presence of private storage, the net public costs of price stabilization increase very rapidly with increasing desired degrees of stability and tend to infinity for perfect price stabilization, the total benefits of such schemes must be evaluated much more carefully. The standard approach of the earlier literature that uses only partial equilibrium producer and consumer surplus measures always gives a finite number of total social benefits. If these are the only social benefits, then, given the results of this paper, a large degree of price stabilization is never socially profitable. This literature disregards the general equilibrium spillover effects of stabilizing one market on others. As Behrman, however, has pointed out, the stabilization of some key markets might have significant impacts on moderating inflation with social benefits much larger than what the surplus measures indicate.

The private trade was shown to have a stabilizing role in the commodity market. Furthermore, the more responsive the private storage community is to intertemporal arbitrage opportunities, the less becomes the necessity of the government to stabilize prices.

The introduction of a futures market was shown to have unambiguously stabilizing effects on interyear price variability. This result is conditional upon the

degree to which the handlers of the physical commodity
make use of the futures market as a hedging medium. A
futures market that is not related to the physical
market does not necessarily have any stabilizing effect.
The job of a public agency in stabilizing year-to-year
prices was shown to become easier in the presence of a
futures market since such a market tends to do part of
the stabilization.

The use of the futures prices by producers as
planning prices was seen also to have stabilizing
influence on the cash prices. An interesting empirical
issue is the extent to which the producers use the
futures prices as instruments for planning. Depending
on the relative tendencies for hedging or speculation
(the magnitudes of h and f in equation (41)), the
futures price can be close to the underlying market
expectation or the current price. In the latter case
the producers are close to having naive expectations,
and cobweb price behavior is not to be ruled out.

The investigation of indirect control of the spot
price via the futures market was seen to pose real
dangers of destabilization and does not seem to be a
viable option, especially in situations where much less
is known about the market than what was assumed here.

Most of the results were shown to apply to
commodities with both lagged and unlagged production.
Furthermore, the form of expectations did not
significantly affect the major results, such as the
influence of private storage and the effects of futures
markets. Given the recent controversy about the
appropriate form of private expectations, it is
reassuring that the results are robust.

The form of price stabilization analyzed was
clearly designed to moderate year-to-year fluctuations.
Nothing was said about within-year price instability.
Many speculative booms and busts belong to this
intrayear variety of price instability and--as has been
shown, for instance, by Salant and Henderson--there
might not be much that a public buffer stock can do in
such a case besides postponing the peak of the boom
or the bust.

It should be clear that public intervention in a
real commodity with the purpose to stabilize yearly
prices is not an easy or a cheap proposition. If the
public agency is less knowledgeable about the
underlying structure of the market than the private
trade, it is bound to do a lot worse than the already
pessimistic results of this paper indicate. Should
the conclusion be that the national and international
commodity markets be left to function uninterfered?
Should we, instead of advocating public storage
intervention, promote the introduction of more
organized futures markets? The answers to these
questions cannot be given without detailed knowledge

of the markets and careful empirical evaluations of alternatives. It is, nevertheless, hoped that the theoretical results of this paper have alerted future empiricists that price stabilization is a far more complicated issue than hitherto assumed.

NOTES

1. For recent surveys of the stabilization literature, see the articles of Turnovsky (1978a) and Wright (1979).
2. In the figure the linear curves do not go through the origin as in equation (3) of the text because in the text we have neglected marginal storage costs. If the marginal storage costs, however, are constant, then the constant term in (3) cancels out from both sides of the market equilibrium relation (4), thus not affecting the analysis.
3. For useful reviews of the literature on the supply of storage theory and its relation to futures markets, see Blau (1944-45) and Weymar (1968).
4. For empirical derivations of optimal buffer stock rules, see the paper by Gustafson (1958) and the more recent works of Pliska (1973) and Gardner (1979).
5. For a good reprint collection of all the major papers on the topic, see the volume edited by Peck (1977).
6. As Anderson and Danthine (1978) point out, equation (36) holds only if the return on futures is stochastically independent of returns on other markets for, otherwise, the market interdependencies might lead to an optimal portfolio such that there is an expected loss from holding futures.
7. McKinnon has also advocated stabilization of the futures markets, not for the objective of price stability but, instead, for income stability of producers.
8. Under the condition $\alpha > b$, the smallest root of (66) is positive and between zero and one.

REFERENCES

Anderson, R. W., and Danthine, J. P. "Hedger Diversity in Futures Markets: Backwardation and the Coordination of Plans." Columbia University Graduate School of Business Research Paper No. 71A. January, 1978.
Behrman, J. R. "International Commodity Agreements: An Evaluation of the UNCTAD Integrated Commodity Programme." Chapter 1 in Policy Alternatives for a New International Economic Order, edited by W. R. Cline. New York: Praeger Publishers for the Overseas Development Council, 1979.

Blau, G. "Some Aspects of the Theory of Futures Trading." Review of Economic Studies. 1 (1944-45): 1-30.

Brennan, M. J. "The Supply of Storage," American Economic Review. 47 (1958): 50-72.

Gardner, B. L. Optimal Stockpiling of Grain. Lexington, Mass.: Lexington Books, 1979.

Gustafson, R. L. "Carryover Levels for Grains." U. S. Department of Agriculture Technical Bulletin No. 1178, 1958.

Helmberger, P., and Weaver, R. "Welfare Implications of Commodity Storage Under Uncertainty." American Journal of Agricultural Economics. 59 (1977): 639-51.

Houthakker, H. Economic Policy for the Farm Sector. Washington, D. C.: American Enterprise Institute for Public Policy Research, 1967.

Hueth, D., and Schmitz, A. "International Trade in Intermediate and Final Goods: Some Welfare Implications of Destabilized Prices." Quarterly Journal of Economics. 86 (1972): 351-65.

Johnson, D. Gale. Forward Prices for Agriculture. Chicago: University of Chicago Press, 1947.

Massell, B. F. "Price Stabilization and Welfare." Quarterly Journal of Economics. 83 (1969): 284-98.

McKinnon, R. "Futures Markets, Buffer Stocks, and Income Stability for Primary Producers." Journal of Political Economy. 75 (1967): 844-61.

Muth, J. F. "Rational Expectations and the Theory of Price Movements." Econometrica. 29 (1961): 315-35.

Newberry, D. M. G., and Stiglitz, J. E. "The Theory of Commodity Price Stabilization Rules: Welfare Impacts and Supply Responses." Economic Journal. 84 (1979): 799-817.

Oi, W. Y. "The Desirability of Price Instability Under Perfect Competition." Econometrica. 29 (1961): 58-64.

Peck, A. "Futures Markets, Supply Response and Price Stability." Quarterly Journal of Economics. 90 (1976): 407-23.

Peck, A. E., editor. Selected Writings on Futures Markets. Readings in Futures Markets. Vol. 2. Chicago: Chicago Board of Trade, 1977.

Pliska, S. R. "Supply of Storage Theory and Commodity Equilibrium Prices with Stochastic Production." American Journal of Agricultural Economics. 55 (1973): 653-58.

Salant, S., and Henderson, D. "Market Anticipations of Government Policies and the Price of Gold." Journal of Political Economy. 86 (1978): 629-48.

Stein, J. L. "The Simultaneous Determination of Spot and Futures Prices." American Economic Review. 51 (1961): 1012-25.

Turnovsky, S. J. "Price Expectations and the Welfare Gains from Price Stabilization." American Journal of Agricultural Economics. 56 (1974): 706-16.

Turnovsky, S. J. "The Distribution of Welfare Gains from Price Stabilization: A Survey of Some Theoretical Issues." In Stabilizing World Commodity Markets, edited by F. G. Adams and S. A. Klein. Lexington, Mass.: Heath-Lexington Books, 1978a.

Turnovsky, S. J. "Stabilization Rules and the Benefits from Price Stabilization." Journal of Public Economics. 9 (1978b): 37-57.

Turnovsky, S. J. "Futures Markets, Private Storage, and Price Stabilization." Journal of Public Economics. 12 (1979): 301-27.

Waugh, F. V. "Does the Consumer Benefit from Price Instability." Quarterly Journal of Economics. 58 (1944): 602-14.

Waymar, F. H. The Dynamics of the World Cocoa Market. Cambridge, Mass.: MIT Press, 1968.

Working, H. "Price Relations Between May and New-Crop Wheat Futures at Chicago." Wheat Studies. 10 (1934): 183-228.

Working, H. "Theory of the Inverse Carrying Charge in Futures Markets." Journal of Farm Economics. 30 (February 1948): 1-28.

Working, H. "The Theory of the Price of Storage." American Economic Review. 39 (1949): 1254-62.

Wright, B. D. "The Effects of Ideal Production Stabilization: A Welfare Analysis Under Rational Behavior." Journal of Political Economy. 87 (1979): 1011-33.

APPENDIX A

Proof that $\partial \sigma_p^2 / (\partial a) < 0$

Considering the market price asymptotic variance relation (12), it is sufficient to show that:

$$\frac{\partial}{\partial a} [1 - r^2] [a + \alpha (1 - r)]^2 > 0 \tag{A1}$$

The derivative in (A1) is equal to:

$$-2r \frac{\partial r}{\partial a} [a + \alpha (1 - r)]^2 + 2 (1 - r^2)$$

$$[a + \alpha (1 - r)] \left(1 - \alpha \frac{\partial r}{\partial a} \right) = 2 [a + \alpha (1 - r)] H_1 \tag{A2}$$

where

$$H_1 = - \frac{\partial r}{\partial a} \{r [a + \alpha (1 - r)] + \alpha (1 - r^2)\}$$

$$+ (1 - r^2). \tag{A3}$$

From Equation (10) that is always satisfied, we obtain:

$$\frac{\partial r}{\partial a} = - \frac{r}{[a + b + 2\alpha (1 - r)]} < 0 \tag{A4}$$

Since the coefficient of $\partial r / \partial a$ in (A3) is negative and the second term in (A3) is positive, then (A4) is sufficient to show that $H_1 > 0$. Hence, (A1) holds.

Proof that $\partial \sigma_p^2 / (\partial \alpha) < 0$

Again, it is sufficient to show that:

$$\frac{\partial}{\partial \alpha} [1 - r^2] [a + \alpha (1 - r)]^2 > 0 \tag{A5}$$

The derivative in (A5) is equal to:

$$-2r \frac{\partial r}{\partial \alpha} [a + \alpha (1 - r)]^2 + 2 (1 - r^2) [a + \alpha (1 - r)]$$

$$\left(1 - r - \alpha \frac{\partial r}{\partial \alpha} \right) = 2 [a + \alpha (1 - r)] H_2 \tag{A6}$$

where

$$H_2 = - \frac{\partial r}{\partial \alpha} \{r [a + \alpha (1 - r)] + \alpha (1 - r^2)\}$$

$$+ (1 - r) (1 - r^2). \tag{A7}$$

From equation (10) again,

$$\frac{\partial r}{\partial \alpha} = \frac{(1 - r)^2}{a + b + 2\alpha(1 - r)} \; . \tag{A8}$$

Substituting (A8) in (A7) and collecting terms, we find:

$$\Pi_2 = \frac{(1 - r)^2 [a + b + rb + \alpha(1 - r)]}{[a + b + 2\alpha(1 - r)]} > 0 \tag{A9}$$

which proves the original assertion.

APPENDIX B

Proof that $\lim\limits_{\alpha \to \infty} \sigma_p^2 = 0$

It is sufficient to show that:

$$\lim_{\alpha \to \infty} (1 - r^2) [a + \alpha(1 - r)]^2 = \infty. \tag{B1}$$

From equation (10), the solution for r is:

$$r = 1 + x - [(1 + x)^2 - 1]^{1/2} \tag{B2}$$

where $x = (a + b)/2\alpha$. Hence,

$$1 - r^2 = 2 - 2(1 + x)^2 + 2(1 + x)$$
$$[(1 + x)^2 - 1]^{1/2}. \tag{B3}$$

$$a + \alpha(1 - r) = a - \alpha x + [\alpha^2(1 + x)^2 - \alpha^2]^{1/2}$$
$$= a - \alpha x + [\alpha^2 x^2 + 2\alpha^2 x]^{1/2}. \tag{B4}$$

Notice that $\alpha x = (a + b)/2 = \text{constant} \equiv c$. Define, also, $d \equiv a - c$.

$$a + \alpha[(1 - r)]^2 = d^2 + c^2 + 2\alpha c + 2d$$
$$(c^2 + 2\alpha c)^{1/2} \tag{B5}$$

Using (B3) and (B5), we have:

$$\frac{1}{2}(1 - r^2)[a + \alpha(1 - r)]^2 = (c^2 + d^2)$$
$$\{1 - (1 + x)^2 + (1 + x)\left[(1 + x)^2 - 1\right]^{1/2}\}$$
$$+ 2\alpha c \{1 - (1 + x)^2 + (1 + x)[(1 + x)^2 - 1]^{1/2}\}$$
$$+ 2d (c^2 + 2\alpha c)^{1/2} \{1 - (1 + x)^2$$
$$+ (1 + x)[(1 + x)^2 - 1]^{1/2}\}. \tag{B6}$$

As $\alpha \to \infty$, $x \to 0$, and the first term in (B6) is finite, the second and third terms can be written after substituting $\alpha = c/x$ as constant multiples of the expression,

$$L(x) = \frac{L_1(x)}{L_2(x)} = \frac{1 - (1 + x)^2 + (1 + x)}{x}$$
$$\frac{}{[(1 + x)^2 - 1]^{1/2}} \qquad (B7)$$

Using l'Hôpital's rule to find the limit of (B7) as $x \to 0$, we obtain:

$$\lim_{x \to 0} L(x) = \lim_{x \to 0} \frac{L_1'(x)}{L_2'(x)} \qquad (B8)$$

$$L_2'(x) = 1 \qquad (B9)$$

$$L_1'(x) = -2 (1 + x) + (1 + x)^2 (x^2 + 2x)^{-1/2}$$
$$+ (x^2 + 2x)^{1/2} = -2 (1 + x) + 2 (x^2 + 2x)^{1/2}$$
$$+ (x^2 + 2x)^{-1/2}. \qquad (B10)$$

It is clear from (B10) that $\lim_{x \to 0} L_1'(x) = \infty$; hence, the original assertion is proven.

APPENDIX C

Proof that $\partial \mu^* / (\partial \alpha) < 0$

From expression (12), it follows that μ^* should be such as to satisfy the relation:

$$\left(1 - r_\mu^2\right)[a + \mu^* + \alpha (1 - r_\mu)]^2 = \frac{\sigma_{\tilde{e}}^2}{\sigma_p^{*2}} \qquad (C1)$$

where the subscript on r denotes the fact that a in equation (10) is replaced by $a + \mu^*$.

Differentiating this expression with respect to α while holding σ_p^{*2} constant, we obtain:

$$-2r_\mu [a + \mu^* + \alpha (1 - r_\mu)]^2 \left(\frac{\partial r_\mu}{\partial \mu^*} \frac{\partial \mu^*}{\partial \alpha} + \frac{\partial r_\mu}{\partial \alpha} \right)$$
$$+ 2 (1 - r_\mu^2) [a + \mu^* + \alpha (1 - r_\mu)] \qquad (C2)$$
$$\left[\frac{\partial \mu^*}{\partial \alpha} + 1 - r_\mu - \alpha \left(\frac{\partial r_\mu}{\partial \mu^*} \frac{\partial \mu^*}{\partial \alpha} + \frac{\partial r_\mu}{\partial \alpha} \right) \right] = 0$$

Solving for $\partial\mu^*/\partial\alpha$, we obtain:

$$\frac{\partial\mu^*}{\partial\alpha} = \left\{ -\frac{\partial r_\mu}{\partial\mu^*} \left[r_\mu\left(a + \mu + \alpha\,(1 - r_\mu)\right) + \alpha(1 - r_\mu^2) \right]\right.$$

$$\left. + (1 - r_\mu^2) \right\}^{-1} \cdot \left\{ \frac{\partial r_\mu}{\partial a} \left[r_\mu\left(a + \mu^* + (1 - r_\mu)\right)\right.\right.$$

$$\left.\left. + \alpha\,(1 - r_\mu^2) \right] - (1 - r_\mu)\,(1 - r_\mu^2) \right\} \tag{C3}$$

From (10), the defining equation for r_μ, we can obtain:

$$\frac{\partial r_\mu}{\partial\mu^*} = -\frac{r_\mu}{a + \mu^* + b + 2\alpha\,(1 - r\,)} < 0 \tag{C4}$$

$$\frac{\partial r_\mu}{\partial\alpha} = \frac{(1 - r_\mu)^2}{a + \mu^* + b + 2\alpha\,(1 - r_\mu)} > 0 \tag{C5}$$

It is clear from (C4) that the denominator of $\partial\mu^*/\partial\alpha$ in (C3) is positive. Substituting (C5) in the expression for the numerator and rearranging terms, we find that the numerator is given by the expression:

$$- [a + \mu^* + b + 2\alpha\,(1 - r_\mu)]^{-1}\,(1 - r_\mu^2)\,\cdot$$

$$[a + \mu^* + b + r_\mu\,b + \alpha\,(1 - r_\mu)] < 0. \tag{C6}$$

Hence, $\partial\mu^*/\partial\alpha < 0$.

APPENDIX D

Proof that $\lim_{\mu\to\infty} R(\mu) = 0$

Define $A(\mu)$ to be the denominator of the expression for $R(\mu)$ in (24). Then apply l'Hôpital's rule to $R(\mu)$.

$$\lim_{\mu\to\infty} R(\mu) = \lim_{\mu\to\infty} \frac{e^{\sigma^2}}{A'(\mu)} \tag{D1}$$

Differentiating $A(\mu)$, we obtain:

$$A'(\mu) = 2 \left[a + \mu + \alpha (1 - r_\mu) \right]$$

$$\left\{ 1 - r_\mu^2 - \frac{\partial r_\mu}{\partial \mu} \left[\alpha (1 - r_\mu^2) + r_\mu \right. \right.$$

$$\left. \left. \left(a + \mu + \alpha (1 - r_\mu) \right) \right] \right\} . \tag{D2}$$

Substituting for $\partial r_\mu / \partial \mu$ the expression in (C4) and rearrange terms we obtain:

$$A'(\mu) = \frac{2 \left[a + \mu + \alpha (1 - r_\mu) \right]}{\left[a + b + \mu + 2\alpha (1 - r_\mu) \right]} \tag{D3}$$

$$\left[a + \mu + b (1 - r_\mu^2) + \alpha (1 - r_\mu) (2 + r_\mu) \right. .$$

As $\mu \to \infty$, $r_\mu \to 0$ (see, for example, Muth (1961)). Since the numerator of $A'(\mu)$ is quadratic in μ (apart from the terms in r_μ) while the denominator is of the first degree in μ, it is clear that $\lim_{\mu \to \infty} A'(\mu) = \infty$ and, hence, that $\lim_{\mu \to \infty} R(\mu) = 0$.

APPENDIX E

Proof of inequality (50)

We have:

$$\text{sign} \left(\frac{\partial \sigma_p^2}{\partial \delta} \right) = \text{sign} \left(\frac{\partial \sigma_p^2}{\partial \theta} \frac{\partial \theta}{\partial \delta} \right) = \text{sign} \left(\frac{\partial \sigma_p^2}{\partial \theta} \right). \tag{E1}$$

From equation (49)

$$\frac{\partial \sigma_p^2}{\partial \theta} = - \frac{\sigma_e^2}{(1 - r^2)^2 \left[a + \eta (1 - r) \right]^4} \frac{\partial}{\partial \theta} (1 - r^2)$$

$$\left[a + \eta (1 - r) \right]^2. \tag{E2}$$

$$\frac{\partial}{\partial \theta} (1 - r^2) \left[a + \eta (1 - r) \right]^2 = -2 \left[a + \eta (1 - r) \right]$$

$$\left\{ r \left[a + \eta (1 - r) \right] + \eta (1 - r^2) \right\} \frac{\partial r}{\partial \theta}. \tag{E3}$$

From equation (48), it is easy to derive the following equation for $\partial r / \partial \theta$:

$$\frac{\partial r}{\partial \theta} = - \frac{b (1 - r)}{a + 2\eta (1 - r) + b (1 - \theta)} < 0 \tag{E4}$$

Combining (E2), (E3), and (E4), inequality (50) is

proven.

APPENDIX F

Proof that $\partial \sigma_{p,\lambda}^2 / \partial \lambda < 0$

From equation (60), we obtain:

$$\frac{\partial \sigma_{p,\lambda}^2}{\partial \lambda} = - \frac{\sigma_e^2}{\left(1 - r_\lambda^2\right)^2 [a + \eta' (1 - r_\lambda) + \rho]^4} \tag{F1}$$

$$\frac{\partial}{\partial \lambda} \left(1 - r_\lambda^2\right) [a + \eta' (1 - r_\lambda) + \rho]^2$$

$$\frac{\partial}{\partial \lambda} \left(1 - r_\lambda^2\right) [a + \eta'(1 - r_\lambda) + \rho]^2$$
$$= -2 [a + \eta' (1 - r_\lambda) + \rho]$$
$$\left\{ \frac{\partial r_\lambda}{\partial \lambda} [r_\lambda (a + \eta'(1 - r_\lambda) + \rho) + \eta' (1 - r_\lambda^2)] \right.$$
$$\left. - \left(1 - r_\lambda^2\right) \left[(1 - r_\lambda) \frac{\partial \eta'}{\partial \lambda} + \frac{\partial \rho}{\partial \lambda}\right] \right\}. \tag{F2}$$

From (55), it can be shown that:

$$\frac{\partial r_\lambda}{\partial \lambda} = \frac{(1 - r_\lambda) \left[(1 - r) \frac{\partial \eta'}{\partial \lambda} + \frac{\partial \rho}{\partial \lambda}\right]}{D} \tag{F3}$$

where

$$D = a + b + 2\eta' (1 - r_\lambda) + \rho > 0. \tag{F4}$$

To prove the original assertion, it is enough to show that the expression in the braces in (F2) is negative. Denote this expression by H_1. Substituting (F3) in H_1, we have:

$$H_1 = \frac{(1 - r_\lambda)}{D} \left[(1 - r_\lambda) \frac{\partial \eta'}{\partial \lambda} + \frac{\partial \rho}{\partial \lambda}\right]$$
$$\left\{ r_\lambda [a + \eta' (1 - r_\lambda) + \rho] + \eta' (1 - r_\lambda^2) \right.$$
$$\left. - (1 + r_\lambda) [a + b + 2\eta' (1 - r_\lambda) + \rho] \right\}. \tag{F5}$$

Denote the braces in (F5) by H_2. By carrying out the multiplications in H_2 and cancelling terms, we obtain:

$$H_2 = - [a + b + \eta' (1 - r_\lambda) + \rho + b\, r_\lambda] < 0 \qquad (F6)$$

Considering the definitions (56) and (57) for η' and ρ, the bracketed expression in (F5) becomes:

$$(1 - r_\lambda)\, \frac{\partial \eta'}{\partial \lambda} + \frac{\partial \rho}{\partial \lambda} = -(1 - r_\lambda)\, \frac{hf}{(h + f + \lambda)^2}$$

$$\pm\ \frac{h\,(h + f)}{(h + f + \lambda)^2}$$

$$= \frac{h^2 + hf\, r_\lambda}{(h + f + \lambda)^2} > 0. \qquad (F7)$$

Equations (F6) and (F7) prove that $H_1 < 0$ and, hence, the original assertion.

4
Monetary Policy
and U.S. Agriculture

John W. Freebairn, Gordon C. Rausser,
and Harry de Gorter

INTRODUCTION

The U. S. agricultural sector has become
increasingly integrated with the domestic and
international macro economies. Policy changes in one
sector can have significant direct and indirect
impacts on the other sectors with concomitant feedback
effects. For example, a restrictive monetary policy
affects general economy wages and prices as well as
interest rates, all of which are cost items in the
production of agricultural products while resulting
exchange rate changes affect commodity demand.
Furthermore, there has been some debate that general
economic policies may have larger effects on the
agricultural sector than traditional commodity
policies. In turn, changes in agricultural prices
affect the general price level and level of economic
activity. Also, there can be effects on the trade
and capital components of the foreign accounts and on
the level of the exchange rate. In this paper, a
small model is built which emphasizes the linkages
between the agricultural sector and the domestic and
international macro-economies and which evaluates the
contributions of policy changes to economic
performance measures. The model is designed for policy
analyses, and a preliminary analysis of a restrictive
monetary policy is reported as an illustrative
application.
In developing the model, three sets of policy
questions were given specific considerations. First,

John W. Freebairn, LaTrobe University, Australia,
Gordon C. Rausser and Harry de Gorter, Agricultural
and Resource Economics, University of California,
Berkeley.

what is the order of magnitude of effects of policy
changes originating in agriculture, the domestic
economy, and the international economy on the three
sectors after consideration of the direct, indirect,
and feedback effects? Second, in evaluating the
performance of the agricultural sector, what is the
relative influence of agricultural commodity policies
and general economy policies? Third, when evaluating
agricultural policies, what weight should be given to
their effects on the rest of the economy and the
balance of payments? Changes in monetary, fiscal, and
exchange rate policies as well as in agricultural
price support, storage subsidy, acreage diversion, and
import quota policies are considered. Nonpolicy
shocks, such as droughts and changes in foreign demand,
can be analyzed. Performance measures include
inflation, wages, and national income for the general
economy; commodity prices and quantities, farm income
and land values for the agricultural sector; and
trade flows, capital flows, and the exchange rate for
the foreign sector. A quarterly simultaneous equation
quantitative model is specified and estimated, based
on models reported in the literature. The task of this
paper is to develop a compact general equilibrium
framework for short and medium term policy analyses.

MODEL FRAMEWORK

To assess and measure the effects of monetary
policy on agriculture, a number of key features and
intersectoral links are incorporated into the model
representation. The model of the general economy has
a fixed price framework following Hicks and Okun.
Aggregate demand is comprised of private consumption
expenditure, private fixed capital investment, change
in inventory, net trade, and government expenditure.
Aggregate supply is represented by price and wage
equations. Nonfarm prices are determined as a markup
over wages (adjusted for productivity) and material
costs. A price expectations augmented Phillips
framework explains wages. These equations provide
the key relationships determining nonfarm prices,
wages, and real income. The general price level, which
also enters the wage equation, is a weighted average
of nonfarm prices and food prices. A conventional
money demand equation is equated to the money supply to
determine interest rates.
The international component is represented by a
balance-of-payments equation including the separation
of the agricultural and non-agricultural trade balance.
Private capital transactions are influenced by relative
interest rates and expected rates of inflation in the
U. S. and in the rest of the world. Prior to 1973, a
fixed exchange rate is specified. Since then,

exchange rates are determined by the supply and demand for foreign currency with an inclusion of exogenous changes in net official reserves. The latter is zero in the case of flexible exchange rates-- the 1981 reference-- and nonzero adjusted by some policy reaction function for a managed or "dirty" float -- the 1973-1979 experience.

The agricultural sector is specified as a flex-price model with a series of supply and demand equations determining price. Agricultural crop production is disaggregated into wheat, coarse grains, and soybeans; other crops are not included. Equations are specified for domestic food demand, export demand, private storage demand, government storage demand, and government export disposal. Planted acreage equations representing planned supply are expressed as functions of expected market prices, government policies such as target prices, loan rates and diversion payments, and input costs. The input costs are related to general economy movements in wages, interest rates, and material costs. Yields are explained by seasonal conditions, technology, current output prices, and current input costs. Livestock products are disaggregated into beef, pork, poultry, eggs, fluid milk, and manufactured milk products. Domestic supply is influenced by expected and past output prices, feed costs, and costs of nonfarm purchased inputs. Allowance is made for cyclical response behavior, particularly in the cattle and hog subsectors. Domestic supply plus government determined import volumes are equated with domestic demand to determine prices. Retail-to-farm price linkage equations are influenced by the costs of nonfarm labor and materials. A set of identities determine income to the crop and livestock activities. The income measure is defined as gross receipts less expenditure on nonfarm inputs and, in the case of livestock, less expenditure on livestock feed.

The distinguishing feature of the model in respect to evaluating the effects of monetary policy on agriculture is the integrative focus that emphasizes intersectoral links and feedback effects between agriculture, the rest of the economy, and the international economy. Until the 1970s, most macroeconometric models and policymakers treated the agricultural sector as exogenous.[1] The experience of high crop and oil prices in the early 1970s induced commercial vendors such as DRI, Chase and Wharton to include agricultural sectors in their macro models (Chen (1977) and Roop and Zeitner (1977)).[2] Current models largely treat agriculture as a satellite system with limited feedback between the general

economy and agriculture. In addition, the models are
very large and are concerned more with forecasting
than with policy analysis. A series of small models
based on a cost markup equation for prices and a price
expectations-augmented Phillips curve for wages were
used in the 1970s to study the effects of increases
in oil and food prices on levels of general prices and
employment under different monetary rules and
expectation models (Okun (1975), Gordon (1975),
Gramlich (1979), Kaldor (1976), Phelps (1968),
Schlagenhauf and Shupp (1978), Van Duyne (1979, 1980),
Lawrence (1980), and Blinder (1979)). These models
ignore feedback effects from the general economy
prices, wages and interest reates on agricultural
supply and on the balance of payments, and they
simplify the dynamic reactions of the crop and
livestock interactions.

Agricultural sector models by Cromarty (1959),
Egbert (1969), and Quance and Tweeten (1972), and
also some more recent ones (i.e., Lamm (1980)), treat
the agricultural sector as a separate sector with
general prices, interest rates, wages and exchange
rates, treated as exogenous. However, the Chen (1977)
and Roop and Zeitner (1977) studies have a more
integrative approach. Recently, a number of models
with an integrated focus between agriculture and the
rest of the economy have been reported, including
Shei (1978), Lamm (1981), and Hughes and Penson
(1980).3/ Lamm's highly aggregated annual model
focuses primarily on real factors and does not
consider the monetary and agricultural sector policy
instruments of interest to this study. The Hughes
and Penson model emphasizes the monetary aspects, but
its annual format and its assumption of flexible
prices in both the general economy and the
agricultural sector detracts from its use for our
purposes. Shei developed a 24-equation annual model
primarily to study the effects of devaluation and to
analyze an exogenous increase in crop export demand.
While the model has most of the important intersectoral
linkages, its high degree of aggregation has meant
that effects of interest rates on private and public
crop storage, most forms of agricultural commodity
policy instruments, and the time dynamics of livestock
cyclical behavior are omitted. Studies of the effects
of exchange rate changes on U. S. agriculture, e.g.,
Bredahl, Meyers, and Collins (1979); Schuh (1974);
Chambers and Just (1979, 1981); and Johnson, Grennes,
and Thursby (1977) while providing part of the story,
ignore any effects on farm input costs. Also, they
treat the exchange rate as exogenous. Recent
theoretical and empirical studies of the exchange rate,
e.g., Branson (1977), Frankel (1979), Kouri and

Porter (1974), and Driskell (1981) indicate the important effect of capital flows which are influenced by monetary policy as well as trade flows on the exchange rate. Our model pays special attention to the interrelationships between money supply, interest rates, capital flows and the exchange rate, and the effects of the exchange rate on crop export demand and prices.

ESTIMATED MODEL

The model is based on behavioral relationships and identities in a quarterly time framework. The model is a simultaneous equation system except for the crop acreage and livestock inventory equations which enter recursively. In this version, the equations are estimated by ordinary least squares using data for 1966 to 1980. This section reports key equations for the agricultural sector, the balance of payments, and the money component.

Agricultural Model

The empirical results for the agricultural sector are reported in this section. Emphasis is placed on the effects of agricultural sector policies and the links to the international economy and the macroeconomy with particular attention given to the impacts of changes in monetary policy.

Crop supply. The crop supply model allows for the effects of changes in variable costs (interest, wages, and material prices) on both yield and acreage decisions. Explicit treatment of governmental policies such as acreage diversion allotments and support prices is given in addition to current and expected market prices, technology, and inclement weather. Table 1 displays the econometric results for wheat, feed grain (corn, oats, barley, and sorghum), and soybeans acreages.[4]
Estimated equations for crop yields shown in Table 2 show that yields are consistently responsive to current and expected market prices and variable costs and to technology and bad weather. The elasticities for each variable in the feed grain and soybean equations are almost identical with each other while the corresponding elasticities are smaller for wheat since a significant proportion is planted in the fall.

Public versus private grain storage. Since government stocks often are 50 percent of total stocks, explicit representation of the interaction between

TABLE 1
Estimates of U. S. Crop Acreage Response Elasticities, 1954-1980

	Effective Support Rate	Effective Diversion Payment	Market Price of Previous Year	Variable Costs of Off-Farm Inputs of Previous year[a]
Wheat				
Short run	.27 (.06)[b]	-.02 (.01)	.15 (.06)	-.15 (.06)
Long run	.57	-.05	.32	-.32
Coarse grains[c]				
Short run	.08 (.05)	-.04 (.01)	.12 (.06)	-.16 (.08)
Long run	.19	-.09	.29	-.38
Soybeans[d]				
Short run	.15 (.06)		.50 (.07)	-.22 (.04)
Long run	.57		1.97	-.85

Source: Computed

[a] Costs of purchased materials, hired labor, and interest payments.
[b] Figures in parentheses are estimated standard errors.
[c] Cross-price elasticities of coarse grains with respect to soybeans are: -.12 (.06) and -.29.
[d] Cross-price elasticities of soybeans with respect to coarse grains are: -.29 (.07) and 1.11.

TABLE 2
Estimates of U. S. Crop Yield Response Elasticities,
1954-1980

	Farm Price	Variable Costs of Off-Farm Inputs[a]
Wheat		
Short run	.09 (.07)[b]	-.11 (.07)
Long run	.09	-.11
Coarse grains		
Short run	.27 (.11)	-.39 (.11)
Long run	.55	-.78
Soybeans		
Short run	.29 (.10)	-.35 (.13)
Long run	.41	-.50

Source: Computed.

[a]Costs of purchased materials, hired labor, and interest payments.
[b]Figures in parentheses are standard errors.

private and public stockholding is paramount. The
government sets loan prices and storage and interest
rate subsidies. The private economy responds by
placing and redeeming grain under the loan and farmer-
owned reserve programs and defaulting to government
stocks. Net of direct government purchases and sales,
governmental inventories are then determined residually.
This can be shown in equation form as:

$$IC(-1) + IPLA = IC + DEF + RED \qquad (1)$$

$$GO(-1) + DEF = GO + GD. \qquad (2)$$

Adding equations (1) and (2) gives:

$$GC(-1) + IPLA = GC + RED + GD. \qquad (3)$$

Subtracting grain redeemed from grain placed under
loan gives net placements allowing equation (3) to be
expressed as:

$$GC = GC(-1) + NP - GD \qquad (4)$$

TABLE 3
Elasticity Estimates of Selected Variables on Net Quantities of Grain Placed Under Government Storage Programs Quarterly, 1966-1980

	Target Price	Market Price	Government Stocks	Interest Rate
Wheat				
Short run	13.4 (3.1)[a]	-13.4 (3.1)	-5.8 (1.7)	.06 (2.1)
Long run	18.4	-18.4	-8.0	.08
Coarse grains				
Short run	16.4 (3.2)	-16.4 (3.2)	-7.6 (1.5)	6.34 (2.01)
Long run	23.4	-23.4	-10.8	9.1

Source: Computed.

[a]Figures in parentheses are standard errors.

where IC = ending stocks under loan (including farmer-
owned reserve), IC(-1) = beginning stocks under loan
(including farmer-owned reserve), IPLA = grain-placed
stocks under loan (including farmer-owned reserve),
DEF = grain defaulted into government stocks, RED =
grain redeemed from loan inventories, GO = ending
government stocks, GO(-1) = beginning government stocks,
GD = net government disposal (sales minus purchases),
GC = total ending stocks under government control (IC
plus GO), GC(-1) = total beginning stocks under
government control (IC(-1) plus GO(-1)), and NP = net
placements of grain under loan (IPLA - RED).

To endogenize the level of government stocks as in
equation (4), equations representing net placements
are estimated with the empirical results given in
Table 3. Changes in beginning public stocks, interest
rates, target prices, and market prices consistently
and significantly affect net placements for both wheat
and feed grains. The level of grain production and
beginning private stocks also significantly and
positively affects net placements but are not
reported in Table 3.

Private stock demand equations are given in
Table 4. Private stocks are affected by production,
government controlled stocks, interest rates, target
and market prices, and incoming private stocks.

Feed demand. The demand for feed is a derived
demand depending on the output of the livestock sector,
a weighted average price of livestock products, and
crop prices. The estimated equations in Table 5
support the derived demand hypothesis.

Crop export demand. Rather than estimate
simplified export demand elasticities and risk the many
biases associated with this procedure (see Orcutt and
Thompson and Abbott), we used Yntema's formulae to
calculate the export demand elasticities as a weighted
sum of domestic demand and supply elasticities in the
principal export-competing and import-competing
countries. We have extended the procedure used by
Bredahl, Meyers, and Collins to allow for time lags in
response and for cross-price effects. Specifically,
for the export-demand elasticity in a particular
period, we use the formulae:

$$E_{ij} = \sum_r \left(\frac{Q_{ir}^d}{Q_i^e} E_{jr}^d E_{ijr}^d - \frac{Q_{ir}^s}{Q_i^e} E_{jr}^s E_{ijr}^s \right) + \sum_t \frac{Q_{it}^m}{Q_i^e} E_{ijt}^m \tag{5}$$

TABLE 4
Elasticity Estimates for Variables Affecting Private Industry Stock Demand for Grains Quarterly, 1966–1980

	Target Price	Government Stocks	Interest Rate	Grain Production	Market Price
Wheat					
Short run	.16 (.07)[a]	-.07 (.02)	-.10 (.07)	.46 (.04)	-.06 (.057)
Long run	.42	-.18	-.26	1.21	-.16
Coarse grains					
Short run	.06 (.047)	-.05 (.024)	-.02 (.036)	.47 (.02)	-.06 (.047)
Long run	.16	-.14	-.05	1.27	-.16
Soybeans					
Short run	0	0	0	.40 (.02)	-.08 (.03)
Long run	0	0	0	1.45	-.29

Source: Computed

[a]Figures in parentheses are standard errors.

TABLE 5
Estimates of U. S. Feed Demand Response Elasticities

	Coarse Grains		Soybeans	
	Short Run	Long Run	Short Run	Long Run
Market price				
Wheat				
Coarse grains	-.13 (.062)[a]	-.21		
Soybeans			-.10 (.036)	-.17
Livestock production index[b]	.14 (.365)	.24	.83 (.382)	1.38
Livestock price index	.14 (.08)	.24	.35 (.074)	.58

Source: Computed.

[a]Estimated standard errors given in parentheses.

[b]Index of beef, pork, milk, poultry, and egg production.

where E_{ij} = elasticity of export demand for i with respect to price; E^d_{ijr}, E^e_{ijr} = elasticity of demand and supply, respectively, for i with respect to price j in country r; E^d_{jr}, E^s_{jr} = elasticity of response of domestic demand price and supply price, respectively, to exporter's price; E^m_{ijt} = elasticity of excess demand in country t (distinct from country r) for imports of i with respect to exporter's price j; Q^e_i = U. S. exports of i; Q^d_{ir}, Q^s_{ir} = quantity of i demand and supplied, respectively, in country r; and Q^m_{it} = net import of i by country t. The preferred strategy was to treat all countries in the first rigght hand term; however, limited data forced the treatment of the centrally planned economies by the second right hand term.

The implementation of the above relationships required data from several sources. Briefly, the country breakdown follows that of the USDA Grain, Oilseeds and Livestock Model (GOL) described by Rojko et al. (a total of 26 country groupings). Quantity data came from USDA publications. The price elasticities refer to three periods: (1) no supply response (one quarter); (2) short-term supply response is allowed (one year); and (3) the long run. The price elasticities of demand, long-run supply, and import demand of the centrally planned economies are essentially those of the GOL model (where available) updated by recent estimates reported in the literature. Values for the price transmission elasticities (with extreme values of zero for completely insulating domestic policies and unity for free trade) are determined arbitrarily after consideration of information on national agricultural policies compiled by FAO, country yearbooks, annual reports of marketing authorities, and professional journal articles. Clearly, some approximations are involved. The derived direct cross-price elasticities of demand for U. S. exports of wheat, coarse grains, and soybeans are given in Table 6.

A shift variable representing the net effect of changes in demand in other countries, e.g., income, changes in supply in other countries, e.g., seasonal conditions, and agricultural policies is determined as a residual. In the present model, these effects are treated as exogenous.

The domestic demand equations. The clasticities for domestic food consumption by commodity group are obtained from the study by George and King. This study used the traditional demand theory constraints to obtain their elasticity estimates. To save space, the parameters are not reproduced here. The George and King commodities have been aggregated to beef, pork,

TABLE 6
Estimated Direct and Cross-Price Elasticities of Export
Demand for U. S. Crops

	Elasticity with Respect to Price of:		
	Wheat	Coarse Grain	Soybeans
One quarter			
Wheat	- .57	.21	0
Coarse grains	.16	-1.00	.45
Soybeans	.02	.08	- .61
One year			
Wheat	-1.34	.47	0
Coarse grains	.40	-1.54	.48
Soybeans	.02	.16	- .94
Long run			
Wheat	-3.80	1.19	0
Coarse grains	1.19	-3.97	.70
Soybeans	.05	.54	-1.76

Source: Computed.

chicken, eggs, fresh milk, manufactured dairy products,
bread and cereal products, fats and oils, and all
others.

Crop demand and prices. Prices for the three
crops are determined by the price which clears market
supply and demand in the identity,

$$Q + GC(-1) + PS(-1) = DF + DFE + DE + DGE + GC$$
$$+ PS + S \qquad (7)$$

where Q is production (acreage times yield), GC
is government stocks, PS is private stocks, DF is food
demand, DFE is feed demand, DE is export demand, DGE is
government exports, and S is seed use.

Dairy sector. The influence of the government
programs on market prices without quantity controls
breaks the usual simultaneity between market prices and
quantities when approached from the perspective that

the government price floors are policy decisions. We exploited this factor by reducing the system to a recursive structure with the influence of Federal policy settings entering the model explicitly. The farm price of milk, PM, was determined as the weighted average of the Class I (fluid milk) and Class II (manufactured milk) prices, P_1 and P_2, as

$$PM = \frac{P_1Q_1 + P_2 (Q - Q_1)}{Q} \tag{7}$$

where Q is total milk production and Q_1 is fluid milk consumption.

Total milk production is the product of the numer of cows milked and the milk per cow. The number of cows milked (CMK) in the dairy sector is affected by the lagged average price of milk (PM) and the price of feed (PF):

$$CMK = 3.8 - .13D1 - .12D2 - .07D3$$
$$+ .071PM(-2) - .0023PF(-2) - .02T$$
$$(.004) \qquad (.0013) \qquad (.006)$$
$$[.17] \qquad [-.06]$$

Milk per cow (MPC) is affected by the relative price of milk to feed (PM/PF), milk per cow lagged, and a trend (T):

$$MPC = 174.4 + 64.4DI - 89.6D2 - 80.1D3 + 103.6\left(\frac{FM}{PF}\right)$$
$$(70.9)$$
$$[.012]$$

$$+ 1.08T + .779 \; MPC(-1)$$
$$(.427) \; (.089)$$
$$[.043]$$

<u>Livestock inventory and output decisions</u>. The inventory investment of breeding stock of hogs (IH) is affected positively by lagged hog prices (PH) and negatively by feed prices (PF) and interest rates (IR):

$$IH = 331.1 + 2501D1 - 268.8D2 - 22.5D3 + 23.1PH(-1)$$
$$(8.2)$$
$$[.11; \; 3.5]$$

$$- 8.3PF(-1) - 18.4IR(-4) + .96IH(-1)$$
$$(5.16) \qquad (34.7) \qquad (.07)$$
$$[-.08; \; -2.4]$$

In contrast, hog production (QH) is negatively related to changes in hog prices and positively related to feed prices and off-farm costs of production (C):

$$QH = 1218.9 - 11.5D1 - 137.7D2 + 282.8D3$$
$$+ .21IH(-3) - 24.4PH + 2.4PF + 80.7C$$
$$(.037) \qquad (3.63) \; (1.46) \qquad (11.2)$$
$$[.48] \qquad [.303] \; [.07] \qquad [.33]$$

Any shock to the economy leading to increased feed prices will exacerbate pig meat prices in the short run through liquidation of hog inventories and increased hog supply. However, hog prices will be higher in the long run compared to initial levels since hog supply will decrease due to the initially depleted breeding stocks.

Similar patterns of cyclical response to price changes are found for cattle and beef. The inventory of beef cows (IBC) is a function of feed and beef prices lagged four quarters [PF(-4) and PBF(-4)] and lagged variable costs of production [CBF(-1)]:

$$IBC = .5698.3 - 16.13DI - 577.4D2 - 549.1D3$$

$$\begin{array}{ccc} - 23.8PF(-4) & + 78.71PBF(-4) & - 120.5CBF(-1) \\ (13.9) & (23.9) & (82.5) \\ [-.06] & [.09] & [-.134] \end{array}$$

Beef cattle slaughter numbers (SBF) are a function of the average of beef cow inventories seven, eight, and nine quarters ago (IBCLA), interest rate (IR), dairy cow inventories [ID(-1)], feed price lagged four quarters [PF(-4)], beef price (PBF), and lagged beef cow inventory [IBC(-1)]:

$$SBF = -257.6 - 260.2DI + 167.7D2 + 480.5D3$$

$$\begin{array}{ccc} + .1456 \ IBCLA & - 135.61R & + .29ID(-1) \\ (.059) & (53.6) & (.69) \\ [.674] & [-.10] & [.47] \end{array}$$

$$\begin{array}{ccc} + 17.23PF(-4) & - 13.99PBF & - 0.28IBC(-1) \\ (3.70) & (-11.9) & (.023) \\ [.20] & [-.09] & [-.129] \end{array}$$

Beef production is determined by the product of slaughter numbers and carcass weight per animal slaughtered. The beef carcass weight (CWT) is a function of two quarter lags on the beef and feed prices and of the carcass weight lagged one quarter:

$$CWT = .142 - .0055DI - .0126D2 - .00275D32$$

$$\begin{array}{ccc} + .000497PBF(-2) & - .000165PF(-2) & + .765CWT(-1) \\ (.000146) & (.0000546) & (.075) \\ [.037] & [.022] & \end{array}$$

Finally, poultry production (QCH) is estimated as a function of lagged broiler price [PBR(-1)], feed-price lagged [PF(-1)], and interest rates (IR):

$$QCH = 1705.5 + 151.4DI + 88.0D2 - 137.4D3$$

$$\begin{array}{ccc} + 38.03PBR(-1) & - 6.93PR(-1) & + 83.77IR \\ (12.07) & (2.40) & (16.68) \\ [.364] & [-.265] & [.225] \end{array}$$

Livestock prices. As for crops, the prices of livestock products are obtained from solution of the identity that supply (domestic production + imports) equals demand. The import quantities of beef and manufactured dairy products are determined as a policy variable. Pork, poultry, and eggs are treated as nontraded goods. In all cases, stock changes are treated as given variables for simplicity.

Domestic price linkages. The degree of interdependence between farm prices, P^f; retail food prices, P^r; nonfarm (product and services) prices, P^n; raw material (principally energy) prices, P^e; general wages, W; and costs of nonfarm inputs (which enter farm supply functions), C^f, in the model is represented by the following set of equations:

Retail-farm price equations:

$$P^r = f(P^f, W, P^e), \tag{8}$$

nonfarm price mark-up equation

$$P^n = f(W, P^e), \tag{9}$$

General price index identity

$$P = \theta P^r + (1 - \theta) P^n, \tag{10}$$

Phillips curve wage equation

$$W = F(EP, Y - YP), \tag{11}$$

and farm purchased input cost identity,

$$C^f = \phi_1 W + \phi_2 P^n + \phi_3 P^e + \phi_4 IR, \tag{12}$$

where EP denotes expected price, Y - YP denotes the difference between realized and potential gross national product, IR denotes nominal interest rate, and all other terms are as defined above. Empirical estimates for the retail-farm price equations are given in Table 7.

Similar to the results reported by Heien and by Lamm and Westcott, our estimates of the retail-farm price equations indicate that changes in farm prices take two quarters before being fully reflected in changes in retail prices. Also, we find no evidence of significant nonsymmetries of response to price rises and falls. In total, equations (8) to (12) describe a highly interrelated process of cause and effect between farm and nonfarm prices.

Links with the International Economy

The general specification of the identity for U.S. transactions with the rest of the world in terms of $U. S. is given by:

$$CX * PC + OX * \frac{PW}{X} - LM * PPL - OM * PW * E + KA$$
$$+ ORT = 0 \tag{13}$$

TABLE 7
Estimates of U. S. Farm-Retail Price Margin Response Elasticities

Consumer Price Index	Ratio of Wages to Labor Productivity Index	Farm Price	Farm Price Lagged One Quarter	Raw Material Prices
Beef	.25 (.06)[a]	.40 (.034)	.36 (.035)	.05 (.044)
Pork	.41 (.069)	.25 (.024)	.28 (.024)	.08 (.051)
Poultry	.10 (.078)	.40 (.032)	.12 (.029)	.17 (.064)
Eggs	.32 (.078)	.79 (.99)	.09 (.036)	-.23 (.06)
Fluid milk	.05 (.057)	.47 (.08)	.21 (.071)	.08 (.032)
Milk products	.19 (.12)	.46 (.12)	-.34 (.091)	.55 (.082)
Fats and oils	2.2 (.48)	.004 (.13)	.26 (.12)	-.69 (.383)
Cereals and bakery products	.011 (.17)	-.10 (.054)	.22 (.042)	.13 (.134)

Source: Computed.

[a] Estimated standard errors given in parentheses.

where CX = real quantity of crop exports, PC = index of crop prices in $U. S., OX = real quantity of other exports, PW = index of world prices (using the same weights as for exchange rate), E = index of exchange rate (defined as number of $U. S. required to purchase a unit of foreign currency) given by the Federal Reserve Board's bilateral 10-country weighted index, LM = real quantity of livestock imports, PPL = index of livestock import prices in $U. S., OM = real quantity of other imports, KA = net change in private capital assets (defined as change in U. S., private assets abroad plus change in foreign private assets in the U. S. less a discrepancy), and ORT = a balancing item representing net change in official assets (by both U. S. and foreign country authorities).

Crop exports, CX, and crop price, PC, are aggregates of the export quantities and prices of wheat, feed grains, and soybeans derived in the agricultural sector. Similarly, livestock imports and prices, LM and PPL, refer to aggregates for beef and manufactured dairy products. Other exports, OX, and other imports, OM, are derived as residuals and are explained in the rest-of-the-economy sector. The net change in private capital assets, KA, is explained by differences in domestic interest and expected inflation rates relative to rest-of-world interest rates and expected inflation.

Different policy scenarios about the exchange rate are handled as follows. Under a fixed exchange rate, official assets, ORT, are endogenous balancing items. With a perfectly flexible exchange rate, ORT ≡ 0 and E is endogenous. A mixed or "dirty" float is modeled by adding a policy reaction function for either E or ORT with the other being endogenous.

Monetary Component

The monetary component of the model and the associated actions of the Federal Reserve Bank's accommodation-nonaccommodation of the government deficit/surplus or sterilization-nonsertilization of foreign reserves are represented as:

$$G - T = \Delta B + \Delta B_{CB} \qquad (14)$$

$$CA + KA + ORT = 0 \qquad (15)$$

$$\Delta B_{CB} - ORT = \Delta MB \qquad (16)$$

where G = government expenditures, T = taxes (net transfers to public sector), and ΔB = net changes in private holdings of government bonds, ΔB_{CB} = net changes in the Federal Reserve Bank's holdings of government bonds, CA = current account, KA = capital account, ORT = official foreign reserve transactions, and ΔMB = change in the monetary base (high-powered

money). Note that G and T are part of the demand model
of the macroeconomy model and CA (= CX * PC + OX * PW/E
- LM * PPL - OM * PW * E), KA and ORT are as defined in
the balance of payments equation (13). Equation (14)
is defined to be the government budget constraint, and
the budget surplus/deficit is financed through bond
purchases/sales by the Treasury. Equation (15)
represents the foreign account or balance of payments
equilibrium under either flexible or fixed exchange
rates. Clearly, official reserve transactions by the
Federal Reserve Bank are larger and more of a factor
under fixed exchange rates. The actions of the Federal
Reserve Bank in (non) accommodating the budget deficit/
surplus or (non) sterilizing foreign reserves is
summarized in equation (16). If the government budget
deficit increases, the Federal Reserve Bank can
accommodate the deficit by purchasing government bonds
from the Treasury and, hence, increase the monetary
base. Likewise, the Federal Reserve Bank can sterilize
an increase in foreign reserves by selling bonds to the
public and, hence, leave the monetary base unaffected.
The money supply is represented by:

$$MS = m \cdot MB, \tag{17}$$

where MS is the money supply (M-2) and m is the money
multiplier. Changes in the money multiplier can result
from Federal Reserve Bank actions (e. g., changes in
commercial bank reserve requirements or in the federal
fund rate), commercial banks behavior (e. g., changes
in lending decisions or in reserves), and the public
behavior (e. g., change in currency demand or in demand
deposits). For simplicity, it is regarded as fixed in
our preliminary work. Changes in the monetary base
result from Federal Reserve Bank behavior discussed
above and summarized in condition (16). The monetary
equilibrium can be summarized by two equations:

$$G - T - ORT = \Delta B + \Delta MB \tag{18}$$

$$MS = L(IR, Y) \tag{19}$$

where L = money demand function, IR = interest rate,
and Y = aggregate private income.
   The interest rate is endogenized through
equilibrium condition (19). Changes in the interest
directly affect private expenditure decisions in the
rest of the economy, grain storage decisions, farm
production costs, and livestock inventory decisions in
the agricultural sector and the capital inflow term of
the balance-of-payments equation.

THE EFFECTS OF CHANGES IN MONETARY POLICY ON U. S.
AGRICULTURE

   Having presented the estimated equations for the

agricultural sector the purpose of this section is to trace the impact on agriculture of the significant change in monetary policy initiated by the Federal Reserve Board in October, 1979, and maintained through the early 1980s in reducing the rate of growth of the money supply by selling government securities to the public. The reduced growth of money and the associated rise in interest rates had a direct effect on the general economy, the agricultural sector, and the balance of payments; and, in turn, there was a sequence of indirect and feedback effects.

For the general economy, the direct effect of the monetary squeeze was to reduce real income, particularly, the fixed investment and inventory components and, in turn, initiate a cycle of downward pressure on rates of growth of wages and nonfarm prices. These price effects were strengthened by price falls for agricultural products and revaluation of the currency as it affects prices of other traded goods including oil. After extended lags, the lower rate of price increase reduced the rate of decline in real money balances and modified pressures on nominal interest rates. Over the longer run, these changes may stimulate real expenditure so that real income returns to, or above, its initial level.

In the agricultural sector, the initial effect of higher interest rates was felt as an increase in production costs and in the costs of holding crop and livestock inventories. The fall in inventory demand and the fall in domestic food demand associated with the drop in real income caused a fall in crop and livestock prices and an initial expansion of crop exports. Since the pressures for exchange rate appreciation were reflected in a reevaluation, the net effect was a further fall in crop prices. The lower output prices reduced producers' intended crop production. In the case of the livestock products, there was a positive force from lower feed costs and a negative effect of lower output prices.

After some quarters, a set of countervailing forces emerged to affect the agricultural sector. First, as wage and nonfarm prices fall in the general economy, the costs of agricultural production tend to fall. Second, the revival of real income stimulates demand, particularly for the more income-elastic livestock products. Third, where the initial response was to reduce production, the forces of supply and demand will raise prices to more attractive levels. In the case of the cattle and hog industries, there will be a process of dampened cyclical adjustment because of the short-term perverse response of supply to increased profitability as inventories of breeding animals are accumulated.

Turning to the balance of payments, the restrictive

monetary policy improved receipts on both the trade
and capital accounts. In the early stages, the fall in
real income and, in later periods, the fall in domestic
prices relative to world prices stimulated exports and
reduced imports. The major short-term response,
however, came from a higher capital inflow attracted by
the rise in interest rates and expected fall in
inflation. If the exchange rate is held fixed, the
increase in official reserves will add to the money
base and undo most of the initial monetary package
unless it is sterilized by the Federal Reserve Board.
At the same time, the buildup of U. S. official
reserves will be counterbalanced by a decline in
official reserves of other countries if they follow a
passive monetary rule or reduce their money supplies.
Such a policy response by foreign countries would have
similar effects as described above for the United
States. However, the exchange rate was allowed to
revalue and induce relative price changes to reduce
exports and increase imports. Any fall in real
interest rates will reduce capital inflows.

In the long run (a period of some years), it is
hypothesized that the effects of the restrictive
monetary policy will fall primarily on nominal variables
and very little on real variables. The rate of
increase of all prices--output prices, input costs, and
wages--will be reduced to around the rate of increase
of the money stock, and there will be minimal changes
in relative prices because of the homogeneity
properties of the supply and demand equations of the
agricultural sector and the price markup and wage
equations of the general economy. Because of the
minimal changes in relative prices, there will be only
small changes in real quantities. But the process of
adjustment to this long-run equilibrium involves a
complex of interaction and feedback effects both within
and between the agricultural, nonagricultural and
foreign economies.

CONCLUDING REMARKS

The dynamic quantitative model presented in this
paper incorporates the interactive and feedback effects
of macroeconomic policies, agricultural sector policies,
and noninstrument shocks on key performance variables
in the agricultural sector, the general economy, and
the balance of payments. Key features of the model
are the explicit and detailed treatment of agricultural
sector policies; explicit treatment of public versus
private grain storage; embedding of monetary and
fiscal policies in a general economy model; explicit
balance-of-payments equation; a flex-price
specification for the agricultural sector; a fixed
price specification for the rest of the domestic

economy; explicit price, cost, and income links between agriculture, the rest of the economy and the rest of the world; money and interest-rate links between agriculture, the rest of the economy and the balance of payments; and dynamic responses of agricultural and nonagricultural investment and inventory behavior. Previous frameworks focusing on agriculture omit some of the intersectoral links, and the perspective offered by macroeconomists has failed to treat the agricultural and food system adequately and to identify the appropriate source of shocks. The present model has integrated equational representations and interrelationships from the literature to form a comprehensive model for policy analyses.

A preliminary illustrative application of the model analyzed the effects of a restrictive monetary policy. Treating the agricultural sector as endogenous and having a balance-of-payment constraint resulted in faster reductions in the rates of general price and wages growth than would have occurred by ignoring these intersectoral effects. Increased interest rates initially caused reductions of crop and livestock inventories, thus adding to supplies and reducing prices. In time the decrease in wages and costs of nonfarm inputs used by agriculture from the general economy sector had favorable longer term effects on agricultural supplies and incomes. The direct and indirect effects of the restrictive monetary policy led to pressure for appreciation of the exchange rate and further price and quantity adjustments. In the very long run, the complex adjustment paths resolved to states of lower rates of growth of prices, costs, wages, and nominal incomes and very small if any changes in real income and quantities.

NOTES

1. The original Brookings model included an agricultural sector constructed by Fox, but it was not included in practice. We do not consider the numerous modeling efforts in which agriculture is treated as one of several distinctive sectors in a large, multisector system based on the input-output system.

2. The Wharton agricultural model, for example, contains 249 equations of the agricultural sector that are to be used in conjunction with the Wharton macroeconomic model. In principle, the two models can be solved simultaneously. Operationally, they are solved iteratively.

3. We do not explicitly review relative studies for other countries such as that by Soe-Lin for Canada and Smith and Smith for Australia.

4. Throughout this section, the terms in round

parentheses ( ) are estimated standard errors, and the
terms in square brackets [ ] are estimated elasticities
at mean values. A comma distinguishes the short- and
long-run elasticities when there is a lagged dependent
explanatory variable.

REFERENCES

Blinder, Alan S. Economic Policy and the Great
    Stagflation. New York: Academic Press, 1979.
Branson, William. "Asset Markets and Relative Prices
    in Exchange Rate Determinations."
    Sozialwissenschaftliche Annalen. 1 (1977): 69-89.
Bredahl, Maury E., William H. Meyers, and Keith J.
    Collins. "The Elasticity of Foreign Demand for
    U. S. Agricultural Products: The Importance of
    the Price Transmission Elasticity." American
    Journal of Agricultural Economics 61 (1979): 58-
    63.
Chambers, Robert G., and Richard E. Just. "Effects of
    Exchange Rate Changes on U. S. Agriculture: A
    Dynamic Analysis." American Journal of
    Agricultural Economics 63 (1981): 32-46.
Chambers, Robert G., and Richard E. Just. "A Critique
    of Exchange Rate Treatment in Agricultural Trade
    Models." American Journal of Agricultural
    Economics 61 (1979): 249-257.
Chen, Dean T. "The Wharton Agricultural Model:
    Structure, Specification, and Some Simulation
    Results." American Journal of Agricultural
    Economics 59 (1977): 107-116.
Cromarty, William A. "An Econometric Model for United
    States Agriculture." Journal American Statistical
    Association 5 (1959): 556-574.
Driskill, Robert A. "Exchange Rate Dynamics: An
    Empirical Investigation." Journal of Political
    Economics 89 (1981): 357-371.
Egbert, Alvin C. "An Aggregative Model of Agriculture:
    Empirical Estimates and Some Policy Implications."
    American Journal of Agricultural Economics 51
    (1969): 71-86.
Fox, Karl A. "A Submodel of the Agricultural Sector."
    The Brookings Quarterly Econometric Model of the
    United States, eds. J. S. Duesenberry et al.
    Amsterdam: North-Holland Publishing Company,
    1965.
Frankel, Jeffrey A. "On the Mark: A Theory of
    Floating Exchange Rates Based on Real Interest
    Differentials." American Economic Review 69
    (1979: 610-622).
George, P. S., and G. A.King. Consumer Demand for
    Food Commodities in the United States with
    Projections for 1980, Giannini Foundation
    Monograph No. 26, Berkeley, 1971.

Gordon, Robert J. "Alternative Responses of Policy to External Shocks." Brookings Papers on Economic Activity. 6 (1975): 183-204.

Gramlich, Edward. "Macroeconomic Policy Responses to Price Shocks." Brookings Papers on Economic Activity, (1979).

Heien, Dale M. "Markup Pricing in a Dynamic Model of the Food Industry." American Journal of Agricultural Economics 62 (1980): 10-18.

Hicks, John R. The Crisis in Keynesian Economics. Oxford: Basil Blackwell, 1974.

Hughes, Dean W., and John B. Penson. Description and Use of a Macroeconomic Model of the U. S. Economy Which Emphasizes Agriculture. Texas A&M University, Department of Agricultural Economics, Departmental Technical Report No. DTR 80-5, 1980.

Johnson, Paul R., Thomas Grennes, and Marie Thursby. "Devaluation, Foreign Trade Controls and Domestic Wheat Prices." American Journal of Agricultural Economics 59 (1977): 619-627.

Kaldor, Nicholas. "Inflation and Recession in the World Economy." Economic Journal 86 (1976): 703-715.

Kouri, Pentti J. K., and Michael G. Porter. "International Capital Flows and Portfolio Equilibrium." Journal of Political Economy 82 (1974): 443-468.

Lamm, R. McFall, Jr. "The Role of Agriculture in the Macroeconomy: A Sectoral Analysis." Applied Economics 12 (1980): 19-35.

Lamm, R. McFall, Jr. Aggregate Food Demand and the Supply of Agricultural Products. U. S. Economics and Statistical Service, Technical Bulletin No. 1656, 1981.

Lamm, R. McFall, Jr., and Paul C. Westcott. "The Effects of Changing Input Costs on Food Prices." American Journal of Agricultural Economics 63 (1981): 187-196.

Lawrence, Robert Z. "Primary Commodities and Asset Markets in a Dualistic Economy." Paper presented at the USDA/Universities Consortium for Agricultural Trade Research Conference on Macroeconomic Linkage to Agricultural Trade. Tucson, Arizona, December, 1980.

Okun, Arthur M. "Inflation: Its Mechanics and Welfare Costs." Brookings Paper on Economic Activities (1975): 351-401.

Orcutt, Guy H. "Measurement of Price Elasticities in International Trade." Review of Economics and Statistics 32 (1950): 117-132.

Phelps, Edmund S. "Commodity-Supply Shock and Full-Employment Monetary Policy." Journal of Money, Credit, and Banking 10 (1978): 206-221.

Quance, Leroy, and Luther Tweeten. "Excess Capacity and Adjustment Potential in U. S. Agriculture." Agriculture Economic Research 24 (1972): 57-66.

Rojko, Anthony, Donald Regier, and Patrick O'Brien, Arthur Coffing, and Linda Bailey. Alternative Futures for World Food in 1985. Volumes 1, 2, and 3. U. S. Economics, Statistics, and Cooperatives Service, Foreign Agricultural Economic Report Nos. 146, 149, and 151, 1978.

Roop, Joseph M., and Randolph H. Zeitner. "Agricultural Activity and the General Economy: Some Macroeconomic Experiments." American Journal of Agricultural Economics 59 (1977): 117-125.

Schlagenhauf, D. E., and F. R. Shupp. "Inflation and Price Controls in a Flexprice-Fixprice Model." Annals of Economics and Social Measure 6 (1978): 501-524.

Schuh, G. Edward. "The Exchange Rate and U. S. Agriculture." American Journal of Agricultural Economics 56 (1974): 1-13.

Shei, Shun-Yi. "The Exchange Rate and United States Agricultural Product Markets: A General Equilibrium Approach." Unpublished Ph.D. dissertation, Purdue University, 1978.

Smith, A. W., and R. L. Smith. "A Model of the Australian Farm Sector: A Progress Report," Economic Record. 52 (1976): 462-82.

Soe-Lin. "A Macro Policy Simulation Model of the Canadian Agricultural Sector," Unpublished Ph.D. dissertation, Carleton University, May, 1980.

Thompson, Robert L., and Phillip C. Abbott. "A Survey of Recent Developments in Agricultural Trade Modeling and Forecasting." New Directions in Econometric Modeling and Forecasting in U. S. Agriculture, ed. Gordon C. Rausser. New York: Elsevier/North-Holland Book Publishing Company, 1982.

Van Duyne, Carl. "The Macroeconomic Effects of Commodity Market Disruptions in Open Economies." Journal of International Economics 9 (1979): 559-582.

Van Duyne, Carl. "Food Prices, Expectations, and Inflation." Williams College, Department of Economics, Research Paper No. 3, Williamstown, Massachusetts, 1980.

Yntema, T. O. A Mathematical Reformulation of the General Theory of International Trade. Chicago: University of Chicago Press, 1932.

Part II

Market Structure and Institutions in International Agricultural Trade

5
Price Supports in the Context of International Trade

Peter Berck and Andrew Schmitz

INTRODUCTION

Price supports have been commonplace in agricultural policy for the last five decades. Their effects have also been analyzed in several studies. For example, Wallace (1962), using a welfare economics framework, concluded that the price support programs proposed by Cochrane and Brannan lead to welfare losses. Omitted from Wallace's analysis, however, are important policy instruments such as acreage controls, and government-held stocks. Also, studies of the effects of price instability and the use of policy instruments to deal with it (for example, the early work by Massell) have major shortcomings in that they only incorporate storage policies as policy instruments; also these studies exclude price uncertainty. Lastly, Schuh (1974), shows how an overvalued exchange-rate influences the need for government policies such as price supports (e.g., an overvalued exchange rate reduces farm income since prices are depressed; hence, the need for price supports). This analysis, however, does not consider how the growth in the export component of U. S. agriculture can affect the choice and effectiveness of domestic policy instruments aimed at supporting farm income.

This paper develops a model which incorporates price supports in an uncertain environment. The international trade sector is incorporated explicitly as are such policy instruments as farmer-held reserves. We demonstrate that the choice of policy changes as the volume which a country exports expands.

Peter Berck and Andrew Schmitz, Department of Agriculture and Resource Economics, University of California, Berkeley.

A THEORETICAL FRAMEWORK

Traditional evaluations of agricultural stabilization programs do not clearly distinguish between decisions taken before and after the state of nature is known. The early work of Oi (1961), notes that expected profits as a function of random prices are always at least as great as profits at expected prices. (Proof: Jansen's inequality and the convexity of the profit function.) Oi's conclusion that stabilization never benefits farmers ignores the sources of agricultural uncertainty and leads to the prediction that farmers have their largest crops when prices are high. By building a simple equilibrium model, Massell (1969) was able to reverse Oi's finding against stabilization programs. Massell argues that there is a supply curve appropriate to each state of nature (and equilibrium prices) and that quantities are determined by the intersection of these ex post supply curves and a linear demand schedule. With only two states of nature, simple geometry suffices to show that government stabilization of price through storage increases "producer surplus" when instability is caused by supply fluctuations. The analysis is flawed by the inability of an ex post supply curve to yield information on true rents or surpluses (e.g., Currie, Murphy, and Schmitz, (1971)). The planting decision-- and hence most of the cost commitment comes before the state of nature is known, and this is what distorts the meaning of an ex post supply curve. A model is needed which treats ex ante decisions separate from ex post outcomes.

By explicitly considering the farmer's planting decisions, it is not difficult to derive the true, or planning, supply curve. Let output Q be a function y of acreage planted A times an ex ante uncertain term ε which represents weather: $Q = y(A) (\varepsilon + 1)$, where $E \varepsilon = 0$ and $\varepsilon > -1$. Expected profit, $E \pi$, as a function of random prices, p, and opportunity cost of land, $r(A)$, is found by solving:

$$\max_{A} E p(\varepsilon + 1) y(A) - r(A)$$

for the maximizer, A^*. The expression $E(p + \varepsilon p)$ is a certainty equivalent or planning price, pp, and it is also expected revenue divided by expected quantity. One can then write the supply problem as the certainty equivalent problem:

$$\max_{A} pp \ y(A) - C(A).$$

The usual marginal conditions yield the true supply curve, $S(pp)$. The remaining task is to find the planning price curve and solve for equilibrium prices and quantities. In the linear demand two states-of-nature case, so often examined in the literature, the planning price curve is simple to derive. Let $p(Q) = a - bQ$ be the demand curve and ϵ take on the values ϵ^* and $-\epsilon^*$ with equal probability. Straightforward calculation gives:

$$pp(Q) = E(p + p\epsilon) = a - b(1 + \epsilon^{*2}) Q.$$

Thus, the planning price curve is always below the linear demand curve. (Just, Hueth and Schmitz, (1982)) give conditions on demand curves for producers to prefer stabilization; these conditions are also sufficient for planning price to be below demand.)

Figure 1 shows a typical equilibrium, where S(pp) intersects $pp(Q)$ at Q^*. Actual prices are P_h and P_L each which occur with equal probability. We will use this model to analyze the U. S. agricultural policy in the post-World War II period.

In the 1950s and 1960s, the price of grain was the government loan rate which was set so high that large stocks accumulated even with export subsidies, acreage limitations, and PL-480 food aid. Many causes are cited for the large stocks including technical change, yield price response, and reduced uncertainty. Analysis of the planning supply curve and planning price curve adds another reason to the list: stabilizing prices with a nonrecourse loan program increased the planning price, and the producers simply responded to that higher price.

The analysis of the loan program requires deriving an appropriate planning price curve. Let P_L be the loan rate. If so little acreage is planted that the market price is always higher than the loan rate, then the planning price curve with loan rate, $pp_L(Q)$, is the same as it was without the loan rate. In symbols, when $Q \leq (a - P_L)/b(1 + \epsilon^*)$, $pp(Q) = pp(Q)$. As soon as the loan rate becomes effective, prices stabilize at P_L because there is (on average) always at least enough released storage to drive prices down to the loan rate in poor crop years and the government supports the price through purchases (at P_L) in good crop years. Figure 2 depicts the equilibrium, Q_L. The government alternately buys and sells $Q_L\epsilon^*$, and the price remains stable at P_L.

Figure 2 also contains the comparison between the no-policy equilibrium, Q_E, and the loan program equilibrium, Q_L. The intersection of the supply and pp curve, shown as a dashed line, is at a lower quantity than the stabilized supported equilibrium, Q_L. Actually, the diagram has been constructed to show a stronger point: the supply pp_L and demand curve have a common intersection so expected storage is zero. Contrary to the results of Massell, the diagram shows

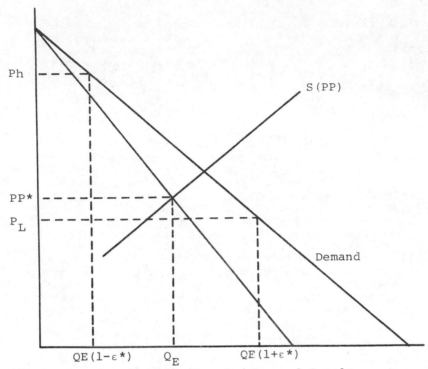

Figure 1   Demand, Planning, Price, and Supply

that, even with linear demand curves, the government cannot stabilize consumption at Q_E and price at $P = a - bQ_E$ by buying and selling $Q_E\varepsilon^*$. The benefits of stabilization engender their own supply response, and equilibrium supply is $Q_L > Q_E$. Alternatively, the government could stabilize price at $pp(Q_E)$, its average. Such price stabilization would require average imports of $Q_E - [a = pp(Q_E)]/b$, again showing that stabilization of price and quantity both at their previous means is impossible.

To sum up, a loan program implies stabilization, stabilization implies an increased planning price, and an increased planning price implies a supply response.

Since the government placed support prices above equilibrium as well as experiencing a stabilized supply response, it had to purchase and hold large stocks. To avoid the costly side effects of the loan program, the government resorted to acreage controls (e.g., 1956), PL-480 giveaways (e.g., 1958) and export subsidies (e.g., 1958). Acreage controls simply shifted the supply curve upwards, and giveaways just avoided the storage (but not the purchase) costs.

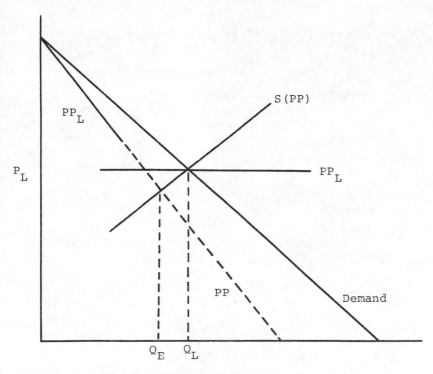

Figure 2 Planning Price with a Loan Program

Export subsidies, however, have real potential for treasury savings. If the demand by foreigners is elastic and grain has already been stored until it is redundant (has zero stock price), then export subsidies can reduce the treasury costs of farm programs.

Even before the 1950s, Secretary Brannan proposed supporting perishable commodities with deficiency payments rather than non-recourse loans. Brannan's plan would have ended the government's practice of buying and destroying potatoes and the specter of the destruction of yet other crops. The deficiency payment idea was widely analyzed and finally implemented in the 1970s following the Russian wheat deal. Deficiency payments alone, however, did not constitute all of farm price policy. A farmer-held-reserve and even a traditional loan policy were used in conjunction with deficiency payments. These policies and the changing world situation lead to a much larger price uncertainty.[1]/ This section describes these policies and their interactions.

DEFICIENCY PAYMENTS

A pure deficiency payment plan consists of setting a target price, P_T, and paying farmers the difference between the market price and the target price whenever the market price is less than the target price. This payment from the treasury is called a deficiency payment. The average revenue curve, $pp_T(Q)$, corresponding to this policy, has three segments. For, when Q is quite low, $Q \leq Q_L$, where:

$$Q_L = \frac{a - P_T}{b(1 + \varepsilon*)}$$

where $pp_T(Q)$ equals $pp(Q)$, the program is never effective. For higher Q, the target prices are effective only some of the time. Letting

$$Q_H = \frac{a - P_T}{b(1 - \varepsilon*)}$$

one finds that for $Q_L < Q \leq QH$,

$$pp_T(Q) = \frac{1}{2}(1 - \varepsilon*) [a - bQ(1 - \varepsilon*)] + \frac{1}{2} P_T (1 + \varepsilon*)$$

which on substituting for P_T gives:

$$pp_T = pp(Q) + 1/2(Q - Q_L) b(1 + \varepsilon*)^2.$$

This regime, where delivery payments are made only in some years, is not covered by Wallace's early analysis. Finally, for $Q > Q_H$, the target prices are always effective and $pp_T(Q) = P_T$ which is precisely the case treated by Wallace. Figure 3 illustrates a pure deficiency payment plan.

Included in the figure as a dashed line is the portion of pp_L not coincident with pp_T. Since pp_T is nowhere below pp_L, producers prefer a target price plan to a loan plan. And, for the same reason, output expands more under a target price plan than under a loan plan.

As Brannan's original critics pointed out, the costs of deficiency payments can far exceed those of loan payments. Assuming that average excess storage is worthless, the costs of a loan plan, c_L, are $P_L (b - a)$ while the costs of deficiency payments are $Q_T(P_T - c')$ where c' is the planning price for Q_T. As the diagram is drawn the latter are larger, but the conclusion could easily be reversed if the diagrm were differently drawn. In particular, commodities such as wheat which are thought of as being in equilibrium on the inelastic portion of the lienar demand curve require deficiency payments on large output quantities.

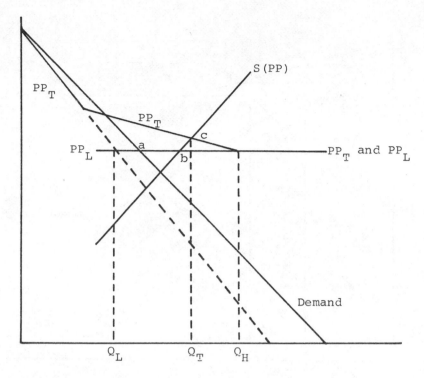

Figure 3   Planning Price with a Deficiency Payment
            Plan

They are poor candidates for a deficiency payment plan.
Butter, on the other hand, would have a higher
elasticity, require subsidization of a smaller quantity,
and the storage would cost relatively more to purchase.
Figure 4 illustrates such a case. Deficiency payments
are P_TCDE which is obviously less than loan payments
FABG. Thus, a deficiency payment plan could involve a
lower treasury cost than a loan plan.

In addition to price-support mechanisms, the
government may subsidize exports. With a loan plan in
place, an export subsidy decreases treasury costs if
and only if it increases revenues from exports. With a
deficiency payment plan in place, an export subsidy may
decrease treasury costs even if it decreases payments
by foreigners.

Consider the simple case where deficiency payments
are made every year. Let $F(P) = f - gP$ be foreign and
$D(P) = a - bP$ be domestic demand. With an export
subsidy of t, the market-clearing domestic price is the
solution of:

$$F(P - t) + D(P) = Q(1 + \varepsilon)$$

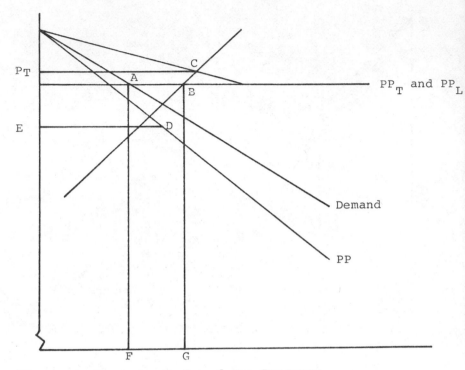

Figure 4   Treasury Costs of Two Programs

for the random value P. In particular,

$$P = \frac{a + f - Q(1 + \varepsilon) + gt}{g + b} .$$

Average treasury costs, C_T, are average farm revenue, $Q_T P_T$, less expected domestic and foreign consumer outlays:

$$C_T = Q_T P_T - E[(p - t) F(p - t) + p D(p)].$$

The incidence of an export subsidy is:

$$\frac{dC_t}{dt} = E \frac{dP}{dt} - 1 F + (P - t) \frac{dF}{dP} + D(P) \frac{dD}{dP} \frac{dP}{dt} .$$

Since dP/dt and dF/dP are nonstochastic and F and D are linear, a certainty equivalent of dC_T/d_t is just the expression in brackets evaluated at $\varepsilon = 0$. Figure 5 shows a target price plan and loan plan in a certainty setting. As drawn, the export subsidy reduces the quantity in the domestic market by $Q_M - Q^*$ which gives a large revenue gain since the elasticity of domestic demand is quite low. The revenue change from the foreign market is nil since the demand elasticity is near unity.

The change in treasury costs is $(P_M - P^*)Q_T - t(Q_T - Q^*)$ and, in this case, it is negative. By way of comparison, the change in the cost of a loan program $P_T(Q_T - Q_L)$, is zero, since at the support price there is no foreign demand.

An exogenous growth in demand, such as that of the early 1970s, reduces the treasury costs of a target price plan even more effectively than do export subsidies. Reinterpreting Figure 5, if the foreign demand curve has expanded by gt, then P* is the new comparison price, and the expected savings are $(P^* - P_M) Q_T$. More generally, an increase in demand decreases treasury costs in all of the models in this paper.

Where treasury costs were the overriding consideration in choosing an export subsidy policy in the 1950s, food security issues have helped shape the policy of the late 1970s. The farmer-held-reserve was seen as a method of stabilizing prices both for American consumers and for foreign customers. In fact, the size of the reserve was to be limited unless the U. S. were to form a commitment to an international grain reserve agreement. This aspect of trade policy was not, as the following section will make clear. in farmers' interests.

With a target-price plan and possibly export subsidies already in effect, producers are not likely to favor further price stabilization. The farmer-held-reserve and the setting of a loan price are examples of further stabilization policy which we will examine.

THE GROWTH IN EXPORT DEMAND

Before proceeding to a rigorous analysis of a reserve policy, the following trade data are important. Table 1 illustrates the growth in U. S. agricultural trade. Between 1965-66 and 1978-79, grain exports in terms of quantity expanded by roughly 10 percent while cotton increased by more than 20 percent. It is apparent that the total demand curve for U. S. agricultural products has been growing (i.e., the combined demand by U. S. consumers and importers has been shifting to the right). Also, partly due to the growth in export demand, the total demand curve facing U. S. producers has become more price elastic (e.g., Burt, Koo, and Dudley, (1980)). As will become apparent, these phenomena have important implications for a combined target-price farmer-held-reserve policy.

The farmer-held-reserve program is a subsidy to the holding of stocks, part of which is withdrawn when the price rises to 170 percent of the loan rate or if

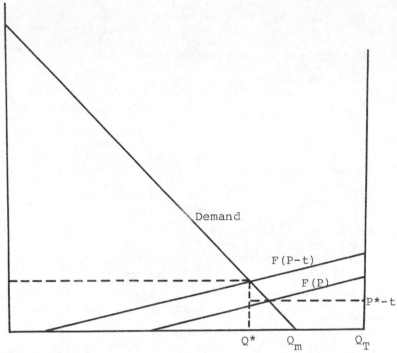

Figure 5   Export Subsidies

TABLE 1
Proportion of Yearly Production of Major U. S.
Agricultural Products Exported (Percent of Total
Quantity), for Selected Years

	1965-66	1970-71	1974-75	1978-79
		percent		
Grains[a]	26.1	20.1	33.4	36.5
Soybeans	28.9	36.9	35.3	38.1
Cotton	31.1	33.7	34.9	53.9[b]

Source: Sarris, Alexander H. and Andrew Schmitz,
"Toward a U. S. Agricultural Export Policy for the
1980s." American Journal of Agricultural Economics.

the farmer elects to sell before the price reaches 140 percent of the loan rate. (Figures are for the 1979 crop year). We will model it simply as a subsidy, t, to be subtracted from the storage costs, c, of holding grain.

Equilibrium response to policy depends on two sets of agents, storers and producers. Producers will behave, as before, maximizing profits without predictive ability. Since producers own land, a factor in fixed supply, they earn rents. Storers, however, are modeled as being in a constant-cost industry without barriers to entry; thus, they earn no profit. For algebraic simplicity, we assume they have perfect information about the next crop before they commit to storage and, further, that bad crops follow good in alternating fashion. This assumption is heroic, but the general conclusions do not change if the world is assumed to have two periods and the storers' have rational expectations.

Without uncertainty in the ordering of the good and poor crops --called instability in the literature--the profits from the storage of crops from one season to the next are:

$$\pi(z_1, Q) = z[a - bQ(1 - \epsilon^*) - bz]$$

$$- z[a - bQ(1 + \epsilon^*) + bz] - z(c - t)$$

where z is the units stored. Since profits are zero in pure competition, the optimal storage, given output Q, will be:

$$z(Q) = Q\epsilon^* - \frac{c - t}{2b}$$

increasing in the government subsidy.

With no policy in effect except storage subsidies, producers clearly gain from the subsidization since:

$$pp_s(Q) = a - bQ - bQ\epsilon^{*2} + bz(Q) \epsilon^*$$

which is their average revenue and is increasing in the government storage subsidy. With a target price also in place, this is no longer the case.

For convenience, assume that storage is not zero over the relevant range of outputs. The planning price curve is as usual composed of three line segments. The first segment of pp_{ST} is relevant when the deficiency payments are made in neither state of nature:

$$Q < Q_L = [a - P_T - 1/2(c - t)]/b$$

and $pp_{ST} = pp_S$. The second segment is appropriate for higher Q:

$$Q_L < Q \le Q_H = [a - P_T + 1/2(c - t)]/b$$

and

$$pp_{ST} = pp_S + 1/2 (Q - Q_L) b(1 + \varepsilon*)^2.$$

The third segment, for $Q > Q_H$ is just P_T. Obviously, pp_{ST} and pp_S are the same when the subsidy, t, is zero. Direct computation shows that the planning price curve is nonincreasing in t:

$$\frac{dpp_{ST}}{dt} = \frac{dpp_S}{dt} - \frac{1}{2} b(1 + \varepsilon*)^2 \frac{dQ_L}{dt}$$

which, on carrying out the algebra, gives:

$$\frac{dpp_{ST}}{dt} = - \frac{1}{4} (1 + \varepsilon*)^2$$

for Q between Q_H and Q_L. Outside this range of Q,

$$\frac{dpp_{ST}}{dt} = 0 \text{ for } Q > Q_H$$

and

$$\frac{dpp_{ST}}{dt} = \frac{dpp_S}{dt} > 0 \text{ for } Q \le Q_L.$$

Thus, in the leading case--that of a target price above market price only some of the time--producers oppose storage subsidies.

PRICE SUPPORTS AND FARMER-HELD RESERVES

To consider the analysis further, the case is presented in Figure 6 where prices fluctuate in the inelastic portion of the demand curve. The stable price is $\bar{P}$, while prices are P_2 and P_1 without storage. Suppose a price support of τ is introduced so producers receive price P_S for Q_1. At P_S, producers prefer price instability to stability (i.e., $\bar{P}$ with storage). Now we introduce a government storage subsidy for producers of the cross-hatched area,[2/] (i.e., $\tau(Q_1 Q_S)$. This storage will be released by producers in period 2 and will cause prices to be P_2' instead of P_2 in that period. Clearly, in period 2, total revenue will decrease due to the release of stocks. In addition, with storage, producers lose P_S abc unless support price P_S is kept. As a result, for producers as a group to participate in the program, either the storage subsidies have to be made greater or a price-support system has to be maintained. Suppose prices continue to be supported at P_S with $Q_S Q_1$ of storage. Clearly, producers are still worse off with

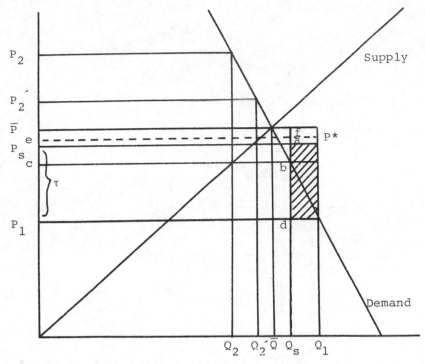

Figure 6   Price Supports and Farmer-Held-Reserves
(inelastic demand)

farmer-held-reserves than with support price P_s and no
reserves. Note that, at P_s with storage by producers,
the Treasury savings are P_1 cbd and P_1 versus P_s with
no storage. But to make storage attractive, the
government has to increase the level of support above
P_s if $\tau(Q_s Q_1)$ is held fixed. There now exists some
price-support level above P_s where instability (and no
stocks) is not preferred to instability accompanied by
supports and farmer-held-reserves.

In the above, even though the support level has to
be raised to induce farmer-held-reserves, there is a
Treasury savings. For example, at a support price P*
producers are better off with storage than with no
storage and P_s. In addition, the Treasury saves cbd
P_1 - efbc. Thus, the government can find a storage
policy for farmers which can both improve their welfare
and reduce Treasury costs. To do this, however, when
reserves are introduced, the government has to increase
the level of price supports even though it reduces the
amount of support payment.

Consider now an interesting comparison in Figure 7
where the demand is price elastic. In the model, τ is

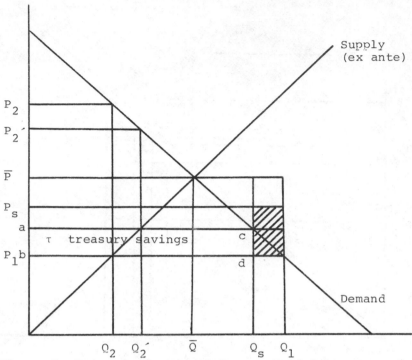

Figure 7 Price Supports and Farmer-Held-Reserves
(elastic demand)

set at the point where producers are indifferent
between a price support of P_s with no storage or
storage and complete stability at P. Suppose farm
storage is subsidized by $Q_1 Q_s(\tau)$. In this case, when
$Q_s Q_1$ is released in period 2, total revenue increases
when prices drop from P_2 to P_2'. As a result, producer
welfare is increased and Treasury costs are reduced by
abdc. In addition, Treasury costs could be reduced
even further before producers would be indifferent
between price supports and no storage or price supports
and storage. In both cases, instability still exists.
There are two important differences between this case
and the inelastic one presented earlier. First, the
Treasury Savings from farmer-held-reserves (keeping
producer welfare unaffected), expressed as a percentage
of total support payments in the absence of reserves,
is greater when the demand is price elastic. Secondly,
in the price elastic case, the level of price supports
does not have to be increased in order to improve
welfare through farmer-held reserves and price supports
above what can be achieved with price supports and no
reserves.

IMPLICATIONS WITH TRADE

Consider the analysis in the above section in the context of the growth of U. S. farm exports through time. The total demand for U. S. farm products not only has grown because of export expansion; it also has become more price elastic. Consider Figure 8 where D_0 is the demand for U. S. products by both domestic and foreign consumers. Prices could be stabilized at $\bar{P}$ with storage. With price supports, producers can prefer price instability to stability by the previous model. Comparing D_0 with D_1, where D_1 represents a different point in time due to the growth in demand from exports, the results of a farmer-held-reserve take on added significance. As already demonstrated, the Treasury savings from a farmer-held-reserve expressed in percentage terms is greater with D_1 than with D_0. This is because, at P*, the demand is price elastic while, at P, it is price inelastic.

Suppose one argued that, with the growth in trade, the variance of instability increases due to weather, exchange rates, etc. (e.g., Schuh (1974) and Schmitz, McCalla, Mitchell and Carter (1981)). In this case, D_1 (with a larger variance in prices than at D_0 since now the price band would be greater than P_1^*, P_2^*) yields an even higher payoff from farmer-held-reserves. As the variance of prices increases, the Treasury savings also increases by introducing both price support and reserves rather than only price supports in order to maintain a certain level of producer welfare.

CONCLUSIONS

In this paper we have modelled agricultural price and income policy through various stages of the development of U. S. agriculture. The emphasis has been on the effects of government intervention in a simple model of uncertainty. We analyzed the interaction effects of price supports and/or deficiency payments, acreage controls, stocks, and export subsidies recognizing the importance of trade. In this context, producers clearly prefer price instability when target prices are used to protect farmers against downside risk. Also, because of the growing importance of international trade, the nature of aggregate demand, treasury costs can be reduded substantially given a specified level at which farm prices are to be supported. Through the use of storage, farm income can be maintained while at the same time governments can reduce their outlays on subsidies.

142

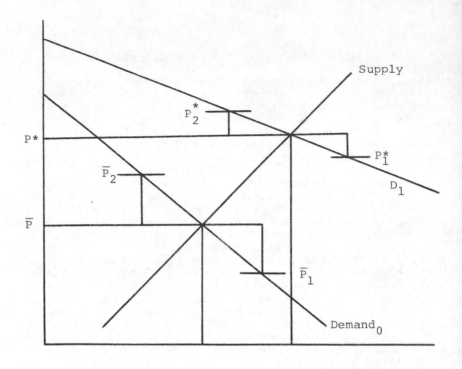

Figure 8   Growth in Export Demand and Price Supports

NOTES

1. Instability is measured by change in value
between present year and past year divided by the value
which is greatest for the two years. An average is
then taken of all percentage changes over the periods
considered. This gives the average degree of
instability. The measure gives a downward bias to
fluctuations and does not correct for trend.

	Corn	Wheat
1949-1955	.071	.043
1956-1969	.059	.078
1970-1979	.176	.166

2. We do not consider here storage by grain
trading firms. Also, we assume that all producers
store in proportion to production. However, these
assumptions could easily be relaxed.

REFERENCES

Burt, Oscar R., Won W. Koo, and Norman L. Dudley. "Optimal Stochastic Control of U. S. Wheat Stocks and Exports." American Journal of Agricultural Economics 62 (1980): 172-87.

Currie, J. Martin, John A. Murphy, and A. Schmitz. "The Concept of Economic Surplus." Economic Journal 81 (1971): 741-99.

Just, R. E., D. L. Hueth and A. Schmitz. Applied Welfare Economics and Public Policy. Prentice Hall Inc., Englewood Cliffs, New Jersey, 1982.

Massell, Benton F. "Price Stabilization and Welfare." Quarterly Journal of Economics 83 (1969): 284-98.

Oi, W. Y. "The Desirability of Price Instability Under Perfect Competition." Econometrica 27 (1961): 58-64.

Sarris, Alexander H., and Andrew Schmitz. "Toward a U. S. Agricultural Export Policy for the 1980s." American Journal of Agricultural Economics, forthcoming.

Schmitz, A., A. F. McCalla, D. Mitchell and C. Carter. Grain Export Cartels. Ballinger Publishing Co., 1981.

Schuh, E. G. "The Exchange Rate and U. S. Agriculture." American Journal of Agricultural Economics, 56 (1974): 1-13.

Wallace, T. D. "Measures of Social Costs of Agricultural Programs." Journal of Farm Economics 44 (1962): 580-94.

6
Pooling Sales Versus Forward Selling and the Management of Risk: The Case of Wheat

Colin A. Carter

INTRODUCTION

International wheat prices have been highly
unstable in the past decade and wheat producers in the
major exporting countries have faced significant amounts
of both price and production risk. The systems of
marketing wheat differ significantly among these
countries as do the methods of risk management by wheat
producers. Australia and Canada have central selling
agencies while in the United States and Argentina the
private trade is responsible for pricing and exporting
wheat.
   Canadian farmers receive a pooled yearly average
price for their wheat and in the United States the
price received is the open market price. However,
through the use of futures and forward contracts, U.S.
farmers can sell their wheat any time before, during or
after the harvest. That is, they can either hedge or
speculate on their wheat sales.
   Some grain produced in the U.S. is sold through
co-operatives, but they do not pool returns from sales
as is done in Canada. An interesting precedent for
grain co-operatives has been set in the U.S. cotton
market however. Calcott, a cotton co-operative, not
only pools sales but also sells forward on the cotton
futures market. This practice of combining pooling
with forward sales has proven to be beneficial to
Calcott's members.
   The objective of this study is to investigate the
merits of the open market versus the orderly marketing
system from the standpoint of price risk management.[1]
The working hypothesis is that the two systems do not

Colin A. Carter, Associate Professor, University of
Manitoba, Winnipeg, Manitoba, Canada

give wheat producers equivalent opportunities to deal
with price uncertainty. Uncertainty arises because
producers must make production decisions without
knowing the eventual output price they will receive.
This paper is written partly in response to the question
--is pooling a substitute for forward contracting or
sales through the futures market?

It is argued that production and marketing decisions
should be made simultaneously, rather than recursively,
by a central selling agency which is interested in
maximizing the welfare of risk-averse producers. This
assumes producers' planting decisions are made ex-ante
the price variability and thus are unable to adjust
their cropping pattern and/or production level once the
state of nature is known. If producers were able to
adjust production levels after the outcome of the
harvest, Schmitz et al. (1981) have shown they may
prefer price stability for some crops but not for
others. The Canadian Wheat Board (C.W.B.) is one
institution which market ex-post production and it could
allow producers to more adequately manage price risk
if it marketed ex-ante production. This would entail
combining pooling with hedging and/or forward sales.
Producers' returns would, however, be left unaffected.

THE CANADIAN SYSTEM OF POOLING AND SELLING

The C.W.B. has the sole authority over both the
export sales and domestic sales for human consumption
of wheat that is produced in the prairie provinces
(Alberta, Saskatchewan and Manitoba). On average, the
C.W.B. markets about 16 million tonnes of wheat per
crop year, or roughly 80 percent of Canadian production.
Revenues from sales within a given crop year (August 1 -
July 31, in Canada) are pooled by grade and each
producer receives the pooled price minus transportation
and storage costs. An initial wheat price is announced
by the C.W.B. but the date of the announcement varies
from year to year. Most often it is announced before
planting. Some years an interim payment is made and
then a final payment is forthcoming six months after
the close of the crop year, when the pooled account is
settled.

The quality of information contained in the initial
price is poor because it is not based on actual C.W.B.
sales. Between 1970-71 and 1979-80, for example, the
initial to total realized price ratio averaged .72 and
this average ratio had a coefficient of variation equal
to .25. By contrast the coefficient of variation of
total realized C.W.B. prices over the same period was
.37. Thus there has been a good deal of price
variability within years under this C.W.B. pooling
system. The reason the producer has poor price

information at planting time on which to base
production decisions is that the pooling of sales and
the transfer of information to producers begins after
production is known. The amount of fertilizer
application, summerfallow acres, and relative acreage
shares among competing crops are all decisions which
critically depend on expected absolute and relative
price levels.

Once the crop is harvested in Canada it is marketed
by the C.W.B. and each producer receives the "average"
price for the year. The timing of sales by producers,
within a crop year, has no effect on the final price
they receive. Exactly when C.W.B. markets the grain
is not public knowledge. However, it is assumed in
this paper that most of it is priced out after
production is known with good certainty (i.e., at the
beginning of the crop year). This corresponds to the
data employed by the Canadian government when it
calculated producer compensation payments arising from
the 1980 U.S.S.R. grain embargo. The government based
its calculations of a $54 million loss to wheat
producers on the assumption that the C.W.B. priced out
9.4 million tonnes of wheat over the first seven months
of 1980. Thus, they assumed sales were made evenly
throughout the crop year.

Export sales of wheat by the C.W.B. are mainly
through direct sales to national trading agencies of
importing countries. Direct sales now account for
between 70 and 80 percent of the C.W.B.'s exports.
However, during the 1950s and 1960s the C.W.B. sold
primarily to the private grain trade. The rising
importance of sales to centrally planned economies and
the declining importance of sales to western Europe
accounted for the major shift away from sales to
intermediaries and towards direct sales.

Almost one-half of Canadaian wheat exports are
currently shipped under Long Term Agreements (L.T.A.s)
with importers. These agreements are signed for
varying lengths of time and normally cover a minimum
volume of trade to be carried out each year during the
agreement. Separate sales contracts are negotiated
under each L.T.A. and pricing is often done on a semi-
annual or quarterly basis. When the L.T.A. is initially
signed, specific details such as grades or prices are
not agreed upon. Once the grain starts moving under
each sales agreement these details are negotiated and
normally a flat price applies to shipments for a three
to six month period, after which the price is
renegotiated.

In sum, the management of price risk by post-
production pooling implies that once a planting
decision is made there is considerable risk that the
final price received will be different from that

prevailing at planting time. This intra-year price
risk is the most important from a producer's standpoint
because inter-year price risk is to a certain extent
avoidable through crop rotation at planting time (Peck,
1975). However, intra-year variability is a real and
unavoidable concern that cannot be dealt with by
altering production plans. The effects of intra-year
price risk on risk averse firms are well known. Feder,
Just and Schmitz (1980) incorporate the possibility of
buying and selling futures into the model of the
competitive firm under price uncertainty. They
theoretically demonstrate that planned production
responds positively to the current futures contract
price and that changes in the subjective distribution
of futures spot prices do not lead to changes in
production decisions. It is demonstrated that a firm
can cope with changes in price uncertainty by
participating in the futures market, where a certain
price can be substituted for an uncertain one, while
optimal production is unaltered. The futures price is
the driving force affecting producer's production
decisions. This result is similarly shown by Danthine
(1978) and Holthausen (1979). Danthine examines the
informational role of futures prices, seen as statistics
by rational traders in formulating their probability
distributions. It is theoretically demonstrated that
with a futures market production decisions are
independent both of the producer's degree of risk
aversion and price expectations, and they are separable
from their "portfolio problem" (from the Markowitz
separation theorem).

REDUCING PRICE RISK VERSUS INCREASING RETURNS: THE DISTRIBUTION OF WHEAT PRICES

The Australian Federal Cabinet has recently
approved changes to the Australian Wheat Marketing Act
which will now allow the Australian Wheat Board (A.W.B.)
to hedge on the U.S. wheat futures market. Dalton (1978)
in a related paper, argues that the A.W.B. could
increase producer returns if it hedged sales on the
futures market. His contention is that the A.W.B.'s
timing of sales has been poor. It is reasoned that the
A.W.B. sells most of its wheat between December and
July and this could be a mistake because the U.S. and
Canada saturate the market before this sales period. He
then proceeds to theoretically show that the distribution
of futures prices should be skewed to the right and that
the futures price will be higher than the actual market
price on more occasions than it is lower. If the mean
of the distribution of wheat prices is higher than the
median price, Dalton points out it would then be
worthwhile for the A.W.B. to hedge wheat sales on the

futures market in order to increase producer returns.

The timing of wheat sales within a particular crop year, once production is known, takes on an importance of its own if Dalton is correct that the probability distribution of wheat prices is skewed. It is therefore worthwhile to empirically investigate the shape of the distribution of wheat prices.

The logarithm of the characteristic function for the symmetric stable family of distribution is:

$$\log_e \psi_p(t) = \log_e \left\{ - \int_{\infty}^{\infty} e^{itp} dF(p) \right\} =$$

$$i\delta t - \gamma \ |t|^{\alpha} = i\delta t - |c \ t|^{\alpha} \tag{1}$$

where it is a real number, and F is the cumulative distribution function (c.d.f.). The characteristic function has three parameters, α, δ, and c. The parameter α is called the characteristic exponent; when $\alpha = 2$ the distribution is stable Gaussian (normal); when $0 < \alpha < 2$ the distribution is stable Paretian; and when $\alpha = 1$ the distribution is Cauchy. The characteristic exponent therefore determines the "type" of a symmetric stable distribution. When α is in the interval $0 < \alpha < 2$ the extreme tails of the stable Paretian distributions are thicker than those of the normal distribution. An important consequence of the stable Paretian distribution is that when $\alpha < 2$, only absolute moments of order less than α exist.

As long as $\alpha > 1$ the mean of the distribution exists and is equal to the location parameter δ. When $\alpha = 2$, the scale parameter, c^{α}, is equal to one-half the variance. Fama and Roll (1968) have suggested procedures for estimating the parameters of the Paretian stable distribution. These procedures are outlined below and the results for the wheat data are reported in Table 1. The random variable analyzed is $Z_t = \ln(P_{t+1}/P_t)$, where P_t represents the average closing Chicago futures price for week t. The data were collected for consecutive weeks within a trading year by contract months.

Given a sample of T observations, Fama and Roll (1968) suggest estimating the scale or dispersion parameter c by:

$$c = \frac{1}{1.654} (\hat{P}_{.72} - \hat{P}_{.28}) \tag{2}$$

where $\hat{p}$ refers to the (f) (T+1) st order statistic, the sample estimate of the f fractile. The standard deviation of the estimator is:

TABLE 1
Parameter Estimates For Symmetric Stable Distributions: Wheat 1966–1976

Chicago Wheat Futures Contract	Truncated Mean ($\hat{\delta}$)	s.e. ($\hat{\delta}$)	Characteristic Exponent ($\hat{\alpha}$)	Scale Parameter ($\hat{c}$)	s.e. ($\hat{c}$)	T
March	-.00052	.00130	1.5	.01436	.00093	539
May	-.00074	.00127	1.4	.01392	.00091	529
July	-.00050	.00128	1.4	.01407	.00091	534
September	-.00011	.00132	1.4	.01455	.00094	541
December	-.00018	.00140	1.5	.01549	.00101	531

Source: Estimated.

$$\sigma(\hat{c}) = 1.5 \ (T^{-1/2}) \ c \tag{3}$$

As an estimator of the location parameter δ, Fama and Roll have suggested using the .5 truncated mean as it has a relatively low sampling variance over a wide range of α's.

For a sample size of T, the .5 truncated mean is:

$$\bar{P}_{.5,T} = \sum_{t=t}^{t_n} \frac{P_t}{(t_u - t_{j+1})} \tag{4}$$

where t_j is the largest integer less than $(.5T/2)+1$ and t_u the largest integer less than $(T-.5T/2)$.

Finally, the following procedure is suggested to estimate the characteristic exponent. First, using some large f (such as f = .95), calculate:

$$\hat{z}_f = \frac{P_f - P_{1-f}}{2\hat{c}} \tag{5}$$

from the sample data. Then an estimate of α can be obtained by searching a table of standardized symmetric stable c.d.f.'s for the value of α whose .95 fractile matches $\hat{z}_{.95}$ most closely. Table 2 in Fama and Roll (1968) can be used for this matching procedure.

Unfortunately the sampling distribution of α is unknown and therefore one cannot determine the level of statistical significance of the estimated characteristic exponent. However, it appears from Table 1 that the empirical distribution deviates from normality since the α's are much smaller than the "normal" value of 2.0. Recall, the "tails" of the empirical distributions are "thicker" the lower the α value.

The estimates of the location parameter (δ) in Table 1 indicates mean price changes over the year average less than twice their standard errors. That is, the average price change is zero.

The function in expression (1) implicitly assumes the empirical distribution is symmetric. Following Roll, (1971) this assumption can be tested by measuring the percentage of sample observations less than the .5 truncated mean. Under the null hypothesis that the distribution is symmetric, the statistic $Z \sim N(0,1)$:

$$z = \frac{(\frac{S}{100} - .5)\sqrt{T}}{.5}$$

where S refers to the number of sample observations less than the .5 truncated mean.

For the empirical distributions in Table 1 the z statistics fall well within the 95 percent acceptance interval and therefore the assumption of symmetry seems appropriate.

The low characteristic exponents reported in Table 1 provide evidence to support the hypothesis that the distributions of wheat futures prices have "thick" tails. This evidence is necessary, but not sufficient, to prove that the empirical distributions were generated by a Paretian stable distribution. The sufficient condition is that the observed empirical distributions were not generated by a mixture of normal distributions with changing variance. Fama and Roll (1971) have suggested a test for choosing between the non-normal stable and the normal distribution with changing variance; one which is based on the property of invariance under addition of a stable distribution. If this property truly holds, then an estimate of α for an entire random sample should be equal to an estimate of α for nonoverlapping sums of observations drawn from the sample. On the other hand, if the sample is drawn from a mixture of normal distributions then the estimate of α should approach 2 as the number of observations in each nonoverlapping sum is increased. Table 2 presents estimates of characteristic exponents for sums of 2, 4 and 6 observations. The results suggest that the true process is non-normal stable for all of the empirical distributions as there is no substantial upward trend in the estimate of α as the number in the sum is increased from 1 to 6.

TABLE 2
Estimates of the Characteristic Exponent for Sums of Observations

Commodity: Contract	Single Observation	Sums of 2	Sums of 4	Sums of 6
Wheat:				
March	1.5	1.4	1.4	1.3
May	1.4	1.4	1.3	1.4
July	1.4	1.4	1.3	1.5
September	1.4	1.5	1.3	1.4
December	1.5	1.4	1.4	1.5

Source: Estimated.

The above results confirm that changes in wheat prices do not follow the normal (Gaussian) probability distribution. It seems evident that the distribution of price changes is leptokurtic (i.e., has greater concentration in the tails than is expected under normality) and that the explanation for the leptokurtosis appears to be that the observations were drawn from stable (Paretian) distributions with infinite variances (see Figure 1). These results show Dalton's reasons for the A.W.B. to trade futures are empirically unfounded. However, it is argued below that there is alternative justification for a central selling agency to hedge on the futures market; namely, to reduce price risk.

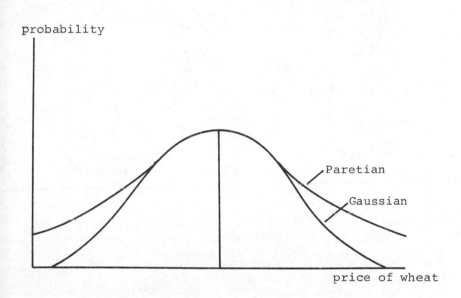

Figure 1   Stable Paretian and Stable Gaussian (normal) Distributions

POOLED PRICES

The above evidence shows that wheat prices have a symmetric stable distribution. Assume further that within a crop year wheat prices have the same expected value[2/] and identical variances. Then, if a group of T farmers (with identical production levels) sell at

different prices during the crop year it is easily shown
that they can pool their returns and reduce the
associated variance while maintaining the same expected
income level, the mean level. This is a generalization
of the central limit theorem. For example, if we have
$R_1, \ldots, R_T$ random returns with the same expected value
and with finite variances $\sigma_1^2, \ldots, \sigma_T^2$

and if, $\bar{R} = \displaystyle\sum_{k=1}^{T} R_k$

then we can pool these returns and each member of the
pool receives:

$$E\left(\frac{1}{T}\sum_{k=1}^{T} R_k\right) = \frac{1}{T}\sum_{k=1}^{T} E(R_k) = \frac{T}{T}\bar{R} = \bar{R} \qquad (7)$$

with variance:

$$\text{Var}(\bar{R}) = \text{Var}\left(\frac{1}{T}\sum_{k=1}^{T} R_k\right) = \frac{1}{T^2}\left[\sum_{k=1}^{T}\text{Var } R_k + \sum\sum_{k=j}\right.$$
$$\left.\text{Cov}(R_k, R_j)\right] \qquad (8)$$

simplifying this expression by assuming independence
we have

$$\text{Var}(\bar{R}) = \frac{1}{2}(T\sigma_k^2) = \frac{\sigma_k^2}{T} \qquad (9)$$

Therefore, price pooling will reduce all uncertainty
associated with expected returns for large T. It is
worth emphasizing that pooling does not affect the
magnitude of the expected return but only its variance.
    The following section compares pooling with the
use of futures or forward contracts. It is shown that
a full hedge or forward sale by a producer after
production is known is equivalent to the pooling result.

HEDGING WITHOUT PRODUCTION RISK

    To analyze the marketing alternatives open to the
producer, besides pooling, let us assume there exists
a wheat futures market which trades contracts calling
for delivery in several time periods throughout the crop
year.[3/] Alternatively, we could assume the existence
of forward contracts for wheat. These contracts allow
producers to market their wheat before it is harvested
or before they physically deliver it. If the futures

market is efficient,4/ its mean price will equal the mean price of the distribution shown in Figure 1, which is also the expected pool price.

Let us denote the beginning of the crop year as period 0 and the end of December as period 1, when the physical sale of the wheat may take place. Without production uncertainty and assuming no basis risk5/the use of futures by the wheat producer has the following expected return:

$$E\ (\tilde{R}) = E\left[\tilde{p}_1 q + \delta(f_0 - \tilde{f}_1)\right] \tag{10}$$

where E is the expectation operator, tildes denote random variables, $\tilde{p}_1$ is the unknown cash price in period 1 (which has a symmetric stable distribution), q is the known output, δ is the proportion of production hedged, f_0 the futures price quoted in period 0, and $\tilde{f}_1$ the unknown futures price in period 1 (which has a symmetric stable distribution). Assuming

$$E\ (\tilde{f}_1) = E\ (\tilde{p}_1)$$

that is, convergence of futures and cash, we have:

$$E\ (\tilde{R}) = E\left[\tilde{p}_1 q + \delta f_0 - \delta \tilde{f}_1\right] = (q - \delta)\ E(\tilde{p}_1) + \delta f_0 \tag{11}$$

with variance

$$\begin{aligned}
\text{Var}\ (\tilde{R}) &= E\ [\tilde{R} - E(\tilde{R})]^2 \\
&= E\ [\tilde{p}_1 q + \delta(f_0 - \tilde{f}_1) - (q - \delta) \\
&\quad E\ (p_1) - \delta f_0]^2 \\
&= E\ [q\ (\tilde{p}_1 - E\ (\tilde{p}_1)\ ) - \delta(\tilde{f}_1 - E\ (\tilde{p}_1))]^2 \\
&= (q - \delta)^2\ \text{Var}\ [\tilde{p}_1 - E\ (\tilde{p}_1)] \tag{12}
\end{aligned}$$

With a full hedge, where $\delta = q$, this variance is reduced to zero. The expected return (11) with a full hedge is equal to that with pooling (7). Thus, pooling is equivalent to fully hedging after production is known.

THE OPTIMAL HEDGE WITH PRODUCTION UNCERTAINTY: CANADIAN WHEAT

Between the time of planting and the beginning of the crop year (before production is known) there is

considerable price risk and pooling after the harvest
does not reduce this risk and its effects on producers.
From 1971 through 1979 the average coefficient of
variation of monthly Chicago cash wheat prices for this
April - November period was .12 and because conventional
pools (e.g., the C.W.B.) market wheat ex-post they do
not reduce this risk faced by producers. This results
in both under-production and resource misallocation.
Futures markets and/or forward sales can reduce some,
but not all, of this risk incurred during the evolution
of the crop year. The element of production uncertainty
and differing attitudes toward risk results in full
hedges being sub-optimal however.

The marketing of Canadian wheat is organized under
the C.W.B. and we will treat producers' income on an
aggregate basis. This section calculates the optimal
hedge for the Canadian wheat crop and relates it to an
unhedged situation. Implications of a non-optimal
hedge on producer output are also drawn. By hedging we
imply either selling short on the futures market or
signing deferred delivery contracts in the spring of
each year. The pooling of returns would still be
practiced.

The model[6]/assumes a two period time frame: pre-
and post-harvest. In the pre-harvest period the crop
is developing and both the post-harvest price $\tilde{p}_1$ and
production $\tilde{q}_1$ are unknown random variables. After the
harvest, production and price are determined. A
futures price f_0, calling for delivery after harvest
is also known prior to harvest.

For a given crop the returns of the producer is a
random variable and is represented as

$$E \ (\tilde{R}) = E \ [\tilde{p}_1 \tilde{q}_1 + \delta (f_0 - \tilde{f}_1)] \tag{13}$$

with variance

$$Var \ (\tilde{R}) = Var \ (\tilde{p}_1 \tilde{q}_1) - 2 \ \delta \ Cov \ (\tilde{f}_1, \ \tilde{p}_1 \tilde{q}_1)$$

$$+ \ \delta^2 \ Var \ (\tilde{f}_1) \tag{14}$$

Assuming producers are expected utility maximizers
and using a mean-variance framework[7]/their objective
function is to maximize

$$E \ (U) = E \ (\tilde{R}) - \lambda \ (Var \ (\tilde{R})) \tag{15}$$

where U is a regular quadratic utility function and λ
is the risk parameter. It can be shown that maximizing
(15) subject to δ gives the optimal hedge as:

$$\delta* = \frac{\text{Cov} (\tilde{p}_1 \cdot \tilde{q}_1, \tilde{f}_1)}{\text{Var} (\tilde{f}_1)} + \frac{f_0 - E (\tilde{f}_1)}{2 \lambda \text{Var} (\tilde{f}_1)} \qquad (16)$$

PRODUCTION AND PRICE RISK

If the C.W.B. were to place a hedge (or make a forward sale) in the spring of the year, equation (16) indicates they would require an estimate of both price and production uncertainty in order to determine $\delta*$, the optimal amount of expected wheat to hedge.

As a measure of production uncertainty this paper uses intended wheat acreage figures published by Statistics Canada (in March) together with a five-year moving average yield and actual production figures. These were collected for fifteen years beginning in 1965. The five-year moving average yield times intended acreage gives a production estimate in March, $\hat{q}_0$. These estimates are then compared with actual production figures released in November, $\tilde{q}_1$. Table 3 reports these data along with annual production errors, defined as $(\tilde{q}_1 - \hat{q}_0) / \hat{q}_0 = \tilde{e}_q$.

Price uncertainty is measured as the difference between the realized December wheat futures price on the first week in December $\tilde{p}_f$, and the December futures price on the first trading day in the previous April, f_0.

That is, the price of December wheat in April is taken as a forecast of the prevailing December price.[8/] Table 4 presents forecast and realized wheat prices with the associated errors, defined as

$$\left(\frac{\tilde{p}_f - f_0}{f_0} \right) = \tilde{e}_p$$

Both of these forecasts were found to be unbiased. The production and price forecast 95 percent confidence intervals are .058 $\pm$ .121 and .140 $\pm$ .270, respectively. Thus, the average forecast errors are not statistically different from zero.

The correlation coefficient between price and quantity uncertainty is -.023. The covariance is -.003. Thus an error in the forecast of price is uncorrelated with an error in the forecast of production. A low price forecast in April tends not to be correlated with a high production forecast in April, and vice versa. In other words, the production uncertainty of the Canadian wheat crop after planting intentions are released, has no significant impact on the world price of wheat. This is understandable as Canada only produces five percent of the world wheat crop.

TABLE 3

Canadian Wheat Production: Forecast and Realized, 1965-1979

Year	5 Year M.A. Yield bu/ac.	Intended Acreage m.a.	Production Forecast m.bu.	Actual Acreage m.a.	Actual Yield bu/ac.	Actual Production m.bu.	Production Error[a]
1965	19.9	28.3	563.2	28.3	24.0	679.2	.206
1966	20.5	29.6	606.8	30.3	27.9	845.4	.393
1967	23.9	31.1	743.3	30.1	19.7	593.0	-.202
1968	23.6	29.4	693.8	29.4	22.1	649.7	-.064
1969	22.8	26.0	592.8	25.0	27.4	685.0	.155
1970	24.2	12.0	290.4	12.5	26.6	332.5	.145
1971	24.7	18.6	459.4	19.2	27.2	522.2	.137
1972	24.6	21.4	526.4	21.3	25.0	532.5	.011
1973	25.7	24.3	624.5	24.8	25.4	629.9	.009
1974	26.3	26.7	702.2	23.5	22.3	524.0	-.254
1975	25.3	25.1	635.0	23.4	26.8	627.1	-.012
1976	25.3	27.6	698.3	27.5	31.4	863.5	.237
1977	26.2	24.7	647.1	25.0	28.9	722.5	.116
1978	27.0	25.5	688.5	26.1	29.7	775.2	.126
1979	27.8	26.8	745.0	25.9	25.1	650.1	-.127

a $\left(\dfrac{\text{Realized-Forecast}}{\text{Forecast}}\right)$

Source: Statistics Canada. Field Crop Reporting. (various issues).

TABLE 4
Wheat Prices: Forecast and Realized
                1965-1979 (¢ per bu.)

Year	C.B.T. December Futures Price on First Trading Day in April	December Futures Price on First Week in December	Forecast Error[a]
1965	149.375	162.062	.085
1966	162.0	178.625	.103
1967	186.00	145.625	-.217
1968	159.5	128.375	-.195
1969	135.5	142.25	.050
1970	143.5	172.25	.200
1971	160.0	170.5	.066
1972	153.5	253.0	.648
1973	216.75	489.5	1.258
1974	399.0	479.5	.202
1975	384.75	341.5	-.112
1976	371.25	256.0	-.310
1977	298.125	266.75	-.105
1978	326.0	374.75	.149
1979	335.75	430.125	.281

[a] $\left(\dfrac{\text{Realized-Forecast}}{\text{Forecast}} \right)$

Source: Wall Street Journal (various issues).

Between price and revenue uncertainty the correlation coefficient is .778 with the covariance being .151. This indicates an underestimate of price in any given year is positively related to an underestimate of revenue, and vice versa.
Substituting

$$\tilde{p}_1 = f_0 (1 + \tilde{e}_p), \quad \tilde{q}_1 = \hat{q}_0 (1 + \tilde{e}_q) \text{ and}$$

$$\tilde{f}_1 = f_0 (1 + \tilde{e}_p)$$

into equation (16), it can be rewritten as

$$\frac{\delta *}{\hat{q}_0} = \frac{\text{Cov} \left[(1 + \tilde{e}_p) (1 + \tilde{e}_q), \tilde{e}_p \right]}{\text{Var} (\tilde{e}_p)}$$

$$- \frac{E (\tilde{e}_p)}{2 \lambda f_0 \tilde{q}_0 \text{Var} (\tilde{e}_p)} \tag{17}$$

The optimal hedge is now expressed as a ratio of forecast output, $\hat{q}_0$.

Table 5 displays the optimal hedge for the Canadian wheat crop. Production uncertainty has little impact on the optimal hedge as it is estimated to be 96 percent of the forecast output for $\lambda > .001$. It is presumed, however, that Canadian producers are sufficiently risk averse such that λ will be larger than .001 and thus close to a full hedge seems appropriate for the C.W.B.

TABLE 5
Optimal Hedge for the Canadian Wheat Crop

Risk Aversion Parameter λ	Average Optimal Hedge as a Percentage of Forecast Output
∞	.96
.001	.96 (.00)[a]
.0001	.92 (.02)
.00001	.55 (.25)

[a]Standard deviations in parenthesis.

Source: Estimated.

CONCLUSIONS

Pooling, as it is now practiced by an agency like the C.W.B., exposes risk-averse producers to a significant amount of price risk between the time of planting and harvest. This risk could be reduced if the pool practiced a hedging (and/or forward selling) program at planting time and thereby passed along a more accurate initial price to producers. This treats production and marketing as simultaneous rather than as recursive decisions. The end result being a larger output and a higher level of utility for producers. It was found that forward selling will reduce price risk but not increase the mean price received.

The optimal hedge, accounting for both price and production uncertainty, was calculated for Canadian wheat producers. It was found that the C.W.B., on behalf of producers, could safely forward sell at least half of the expected output each spring before planting. The total Canadian wheat crop is less than three percent of the annual volume on the Chicago Board of Trade and a C.W.B. hedge or forward sale should not depress wheat prices drastically. Importers could also reduce price risk through forward purchases from central selling agencies directly or on the futures market.

These somewhat strong results arise from the assumption that producers base their cropping patterns and production levels on relative price expectations. Once a decision is made, the producer routinely attempts to minimize the variance associated with the expected return by selling forward. For producers who have zero, or few cropping alternatives, this routine hedging strategy may be sub-optimal. Further research is required to develop optimal selective hedging strategies for these producers.

NOTES

1. For a comparison of other characteristics of the two systems see McCalla and Schmitz.
2. This assumption seems appropriate given the results in Table 1.
3. The Chicago Board of Trade offers wheat contracts expiring in the months of July, September, December, March and May.
4. The pricing efficiency of the wheat market has been the subject of numerous studies. Although somewhat inconclusive, they have found the wheat market to be relatively efficient.
5. The additional basis risk in wheat was found to be relatively unimportant by Nelson. Furthermore, basis risk is not an issue when forward contracting is practiced.
6. See Rolfo for a similar framework.
7. Mean-variance analysis requires a quadratic utility function which has the property of increasing

absolute risk aversion. This has received considerable
criticism in the literature but has been defended for
pragmatic reasons.

    8. Just and Rausser have found the wheat futures
price to be a relatively good forecast.

REFERENCES

Bray, C.E., P.L. Paarlberg and F.D. Holland. The
    Implications of Establishing a U.S. Wheat Board,
    Foreign Agricultural Economic Report Number 163,
    United States Department of Agriculture, April 1981.
Dalton, M.E. "Benefits and Costs of a National Marketing
    Authority Hedging with Production Uncertainty."
    International Futures Trading Seminar: Proceedings
    (1978), Vol. 5, Chicago Board of Trade: Chicago.
Danthine, J.P. "Information, Futures Prices, and
    Stabilizing Speculation." Journal of Economic
    Theory, 17, No. 1 (1978): 79-98.
Fama, E. and R. Roll. "Some Properties of Symmetric
    Stable Distributions." Journal of the American
    Statistical Association, 63, No. 323 (1968): 817-836.
    _____. "Parameter Estimates for Symmetric Stable
    Distributions." Journal of the American Statistical
    Association, 66, No. 334 (1971): 331-338.
Feder, G., R.E. Just and A. Schmitz. "Futures Markets
    and the Theory of the Firm Under Price Uncertainty."
    Quarterly Journal of Economics, 94, No. 2 (1980):
    317-328.
Holthausen, D.M. "Hedging and the Competitive Firm
    Under Price Uncertainty." American Economic Review,
    69, No. 5 (1979): 989-995.
Just, R.E. and G.C. Rausser. "Commodity Price
    Forecasting With Large-Scale Econometric Models
    and the Futures Market," American Journal of
    Agricultural Economics, 63, No. 2 (1981): 197-208.
McCalla, A.F. and A. Schmitz. "Grain Marketing Systems:
    The Case of the United States versus Canada."
    American Journal of Agricultural Economics, 61,
    No. 2 (1979): 199-212.
Nelson, R.D. Should Commercial Grain Farmers Use
    Futures Markets to Hedge Against Price Risk?
    Unpublished, Ph.D. dissertation, U.C. Berkeley,
    1981.
Peck, A. "Hedging and Income Stability: Concepts,
    Implications, and an Example." American Journal
    of Agricultural Economics, 57, No. 3 (1975): 410-
    419.
Rolfo, J. "Optimal Hedging Under Price and Quantity
    Uncertainty: The Case of a Cocoa Producer."
    Journal of Political Economy, 88, No. 1 (1980):
    100-116.

Roll, R. The Behavior of Interest Rates: An
     Application of the Efficient Market Model to U.S.
     Treasury Bills, New York: Basic Books, Inc., 1971.
Schmitz, A., H. Shalit, and S.J. Turnovsky. "Producer
     Welfare and the Preference for Price Stability."
     American Journal of Agricultural Economics, 63,
     No. 1 (1981): 157-160.

7
The U.S. Agricultural Trade Issue Decision Process: An Illustrated Partial Anatomy

George E. Rossmiller

INTRODUCTION

The newspaper headline reads "The USDA ANNOUNCED
..." or the U.S. GOVERNMENT POSITION WAS REVEALED TODAY
ON ..." or "GATT CONFEREES AFTER LONG DELIBERATION
CONCLUDED ...". In each of these cases a decision has
been made and is being communicated to the public. The
decision being reported was not that of an agency, an
office, or an international organization even though
the Secretary, the President or the Director General
may have articulated it to the public. More than
likely, behind the announcement, it was a consensus
decision of a diverse group of individuals who began
with widely differing biases, views, philosophic
perspectives and "official" positions on the particular
issue. Thus the decision was most likely taken after
long, sometimes tedious meetings and debate at various
levels, often accompanied by the passing of mountains
of memos, briefs, options and position papers, some
with sound and useful analysis and argument and some
with no relevant content and perhaps no purpose except
to confuse, delay or divert the focus from the main
issue.

Management, decision theory and political science
texts contain discussions and models of the decision
making process. Various paradigms are used but all
generally include an orderly and iterative set of steps
that usually begins with identification of a problem,
move to collection of relevant data and information,
which in turn is used for analysis leading to
development of alternative courses of action for
problem solution. These alternatives, presumably clear
cut and definitive, are then presented to the decision

George E. Rossmiller, Director, Planning and Analysis
Office of the Administrator, FAS, USDA, Washington,
D. C.

maker who chooses one for action. It is implemented
and then monitored to determine if the problem has been
solved. If the old problem remains or new ones are
identified the process is repeated. While, in general,
such models are correct, the extent of their
abstraction and simplification is such that they fail
to fully account for the complexity and the amount of
iteration that takes place during the decision process.

This paper has modest objectives. It is not an
attempt to improve on existing models of decision
making. Rather it merely describes the system and
mechanisms available to federal executive branch
agricultural decision makers in the narrow field of
trade policy. A case example of the 1980 canned
mushroom import relief decision is used to illustrate
the workings of the system. Hopefully the case
illustration and the conclusions drawn will provide
insights for trade researchers and analysts as to how
their work can be more relevant to the trade policy
decision process. 1/

U.S. TRADE POLICY DECISION SYSTEM

Within the USDA, at the level of the Office of the
Secretary, responsibility for trade policy matters
falls to the Undersecretary for International Affairs
and Commodity Programs. The Foreign Agricultural
Service (FAS) is the operational agency responsible to
this Undersecretary that takes the lead in dealing with
agricultural trade issues.

Within FAS, commodity information and expertise is
found in the six commodity divisions under the
Assistant Administrator for Commodity and Marketing
Programs (CMP), while information and expertise on the
trade policy and border protection regimes of foreign
countries and on the international trade rules of the
game are found in the two regional divisions under the
Assistant Administrator for International Trade Policy
(ITP). International Trade Policy has the
responsibility at the working level for USDA liaison
with the other agencies of government on trade issues.

The Trade Agreements Act of 1979, in addition to
establishing the legal basis for U.S. compliance with
the agreements resulting from the Tokyo Round of
multilateral trade negotiations (MTN), provided the
mandate for a reorganization of the trade functions
within the U.S. Government. An Executive Order, signed
by President Carter on 2 January 1980, put the
reorganization into effect. Simply stated, the
reorganization placed the responsibility of trade
policy formulation in the hands of the Office of the
U.S. Trade Representative and the responsibility for
nonagricultural trade policy implementation in the

hands of an expanded Department of Commerce.
Agricultural trade policy implementation was left with
the U.S. Department of Agriculture with FAS retaining
the lead role.

Several interagency bodies carry out the
coordination of trade matters. The Trade Policy
Committee (TPC) provides interagency coordination at
the cabinet level with the U.S. Trade Representative
(USTR) as Chairman. Subsidiary bodies of the TPC are
the Trade Policy Review Group (TPRG) at the Assistant
Secretary level and the Trade Policy Staff Committee
(TPSC) at the working level. Most trade issues are
country and commodity specific and these are the types
of issues that are the focus of this paper. As can be
seen from a scanning of the list of subcommittes and
taskforces of the TPSC in the appendix, however, the
interagency process is organized to also take up
generic issues and general trade policy matters such as
agricultural commodity agreements, or how to deal with
nonmarket economies, or is there a role for cartels, or
should the U.S. subsidize exports. In 1980, the TPSC
had seven geographic subcommittees, 25 functional
subcommittees and eight task forces at the technical
working level.

Agency participation in these various interagency
groups varies depending upon the subject, the issue,
and most importantly, the constituency they represent,
but the main actors on most agricultural trade matters
include, in addition to USDA and USTR, the Departments
of State, Commerce, Labor, and Treasury and
occasionally the Council of Economic Advisors (CEA).

Trade problems are identified through three main
sources. U.S. government officials may identify an
issue or potential issue during the normal course of
their activities. An example is the need for Section
22 2/ action when market conditions of a particular
commodity covered by Section 22 reach predetermined
trigger point levels or when commodities are known to
be sensitive but are eligible to be placed on the
Generalized System of Preferences (GSP) 3/ list in
periodic GSP list reviews.

Foreign governments may bring trade issues to the
attention of U.S. government officials. For example,
U.S. exports may not be conforming to the foreign
government's health and sanitation regulations as was
the case in Japan when the Mediterranean fruit fly
infestation occurred in California; or the U.S. may be
accused of not living up to MTN concessions as was the
issue when Canada pushed its interest in exporting live
cattle to the U.S.; or a foreign country may expect
injury to its export market through changes in U.S.
legislation or policy as in the case of Australia when

the U.S. changed its Meat Import Law to use a counter cyclical formula for determining allowable meat import levels.

U.S. producer and trade interests may also bring trade issues to the attention of U.S. government officials. In many instances this is done informally during the normal course of contact between the private sector and government. Such issues usually concern actions or contemplated actions by the foreign trade, the foreign government or perhaps the U.S. Government that are outside established policy or guidelines and that will change the existing status quo. These informal contacts may lead to a more formal action on the part of the private sector. Formal action normally consists of filing a petition under a specific authority in U.S. trade legislation alleging injury or the threat of injury due to the action of a government or foreign trade interest.

Agricultural trade policy issues are of three main types. First is the maintaining or expansion of U.S. access in importer markets. Often importing countries take actions that may limit U.S. access and/or impair a concession that they had made under previous GATT agreements. A case in point is the attempted redefinition by the Japanese of high quality beef under the quota they had agreed to provide for access of U.S. high quality beef into their market. The redefinition would have allowed Australian beef to qualify and thus Australia to fill part of the quota that the U.S. had negotiated and paid for with balancing concessions to the Japanese during the MTN. Another case in point is the recurrent threat by the European Community (EC) to impose a consumer tax on vegetable oils, an action that would impair the zero tariff GATT binding on U.S. soybean exports to the EC.

Second is the actions taken by export competitors of the U.S. that cause injury or threaten injury to U.S. exports into third markets. An example is the wheat flour GATT case now pending against the EC in which the U.S. argued that the EC subsidies on wheat flour allowed them to capture an increasing share of the flour market both world wide and in specific flour import markets by undercutting U.S. exporter offer prices.

Third is the actions taken by exporters to the U.S. that cause or threaten to cause injury to U.S. producers in the domestic market or that impair or threaten to impair the ability of the U.S. Government to carry out effectively a domestic farm program. An example of the former is the mushroom import issue discussed in greater detail below and of the latter, Section 22 actions on such commodities as dairy products, sugar or peanuts.

It should be instructive of the trade issue decision process to describe in some detail a case example. The case chosen is a domestic injury complaint by the Pennsylvania mushroom industry requesting escape clause action by the U.S. Government to provide temporary relief from import competition. Historically, U.S. trade legislation has provided for such relief petitions. The case illustrates that normally trade issues tend to have a long gestation period and to be very difficult to resolve.

THE CANNED MUSHROOM CASE

The domestic mushroom industry has had a long history of seeking import relief whenever they have perceived that imports have affected domestic prices and market volumes. In 1964 in response to a petition filed by the domestic canners, the Tariff Commission conducted an investigation resulting in the denial of import relief; in 1966 an industry request for negotiations with Taiwan and Korea to limit mushroom exports was denied. In 1972, after another Tariff Commission investigation, discussions held with Taiwan and Korea on unilateral export restraints resulted in no agreement.

In September 1975 the domestic industry filed a complaint under Section 201 of the Trade Act of 1974 4/. The U.S. International Trade Commission (USITC) found serious injury and recommended expedited adjustment assistance. A sharp increase in mushroom imports occurred between June and August 1976. The President directed USTR to ask USITC to do an expedited Section 201 investigation and to seek assurances from Taiwan and Korea that they would regulate the flow of mushroom exports to prevent a surge during 1977. Assurances were given in September 1976. The normal procedure when voluntary restraint of exports is sought is for the governments to agree on an annual volume of export and for the government of the exporting country to carry out the agreement through the use of export licenses.

The USITC delivered a serious injury finding in March 1977, whereupon President Carter denied import relief but asked USITC to monitor the import situation. Assurances were again sought and given that no surge of exports would occur from Taiwan and Korea in 1978. Taiwan announced its intent to restrain exports through November 1979 and Korea appeared to exercise restraint into 1980. In the meantime exports by Hong Kong grew rapidly, moving from 1.0 million pounds in 1976/77, to 7.4 million pounds in 1977/78, to 12.6 million pounds in 1978/79 and to 17.6 million pounds in 1979/80 (Table 1).

An apparent import surge began during the last
half of calendar 1979 and continued into 1980. Imports
were up 22 percent in the first quarter of 1980
compared to the same period in 1979, when imports had
been off. With the end of Taiwanese restraints at the
end of November 1979 imports from Taiwan increased
sharply to 76 percent above first quarter 1979 levels.
On a trend basis, however, total first quarter 1980
imports were slightly below what might have been
expected, 25 million pounds actual versus 25.7 million
pounds projected trend. Thus the apparent import surge
in late 1979 and in early 1980 was mainly a catching up
to the projected trend. The first arrivals of
mushrooms from the Peoples Republic of China (PRC)
occurred in March 1980.

TABLE 1
Canned Mushroom Imports by U.S., Marketing Years
1973-74 to 1979-80 (Million Pounds Drained Weight
Basis)

	Taiwan	Korea	Hong Kong	Other	Total
1973-74	30.4	10.3	0.2	4.6	45.5
1974-75	34.2	11.0	0.1	4.7	50.2
1975-76	36.1	18.0	0.1	3.1	57.4
1976-77	43.0	22.2	1.0	3.3	69.4
1977-78	57.2	23.7	7.4	3.7	92.0
1978-79	42.1	27.0	12.6	4.5	86.2
1979-80	61.3	25.9	17.6	8.8	113.6

Source: Table 9, USITC Report No. TA-201-43, p. A-78.

Nevertheless this apparent surge in imports was
coupled with declining prices during the first quarter
of 1980. Between December 1, 1979 and March 1, 1980 the
price of Number 1 mushrooms for canning dropped from 62
cents per pound to 47 cents; Number 2 mushrooms for
canning dropped from 51 cents per pound to 47 cents per
pound; and utility quality mushrooms rose slightly from
41 cents per pound to 42 cents per pound. On March 14,
1980, the American Mushroom Institute filed a petition
for import relief under Section 201 of the Trade Act of
1974. The USITC initiated its investigation on March
24, 1980.
On March 31, 1980, only 7 days after the USITC
began its investigation, Senators Heinz and Schweiker
and Congressman Schulze of Pennsylvania sent a telegram
to President Carter indicating that mushroom prices
were below the cost of production and stating that
unless immediate relief action was taken, the domestic

mushroom industry would be extinct before the USITC investigation was completed. They asked that an interagency task force be convened to address the short-term relief issue and that the USTR negotiate with Taiwan, Korea, Hong Kong and the PRC to restrain exports to the U.S. until the domestic mushroom price recovered to cost of production levels.

The USDA, primarily the Horticultural and Tropical Products Division and the International Trade Policy area of FAS began compiling the information available and the questions to be answered before an agency position could be developed. Several facts of importance were revealed in the early stages of the USDA inquiry. First, the U.S. mushroom industry as a whole had experienced very rapid growth over the 1968/69-1978/79 decade with production increasing by almost 2-1/2 times and farm value of production increasing from $68 million to $362 million during the period. (Table 2) Growth occurred from major increases in fresh mushroom consumption as well as lesser increases in processed mushroom consumption. The Pennsylvania mushroom industry heavily oriented to canned mushroom production, showed significantly slower growth during the period relative to other producing areas.

Further, the industry was undergoing rapid concentration. During the 1969-1979 period the number of growers declined by 23 percent and the number of canners declined by more than a third. By 1979, three percent of the growers accounted for 50 percent of total mushroom production. The larger producers and canners have the means to adopt new technology to increase productivity while the smaller ones do not. Thus the smaller firms have found themselves in a decreasingly competitive position within the domestic industry. The smaller firms are concentrated largely in Pennsylvania.

One cause for the increased volume of mushroom exports to the U.S. was the operation of a licensing system in the EC that restricted entry of mushrooms into the EC market. (Table 3) West Germany, the major EC mushroom market did not import any mushrooms in 1979 from Taiwan, Korea or Hong Kong because of the denial of import licenses. The U.S. and Canada were presumably receiving mushrooms that normally would have gone into the EC market. During regularly scheduled semi-annual U.S.-European Community consultations in May 1980 the U.S. raised the question with the EC of whether they might relax their mushroom import restrictions. The EC responded that they had already decided to do so effective May 15, 1980. This action was expected to relieve some of the pressure of exports to the United States.

TABLE 2
Volume, Price and Value of U.S. Mushroom Production, 1968-69 to 1979-80

Year	Fresh Market Volume million lbs.	Fresh Market Price cents per lb.	Fresh Market Value $ million	Processing Volume million lbs.	Processing Price cents per lb.	Processing Value $ million	Total Volume million lbs.	Total Price cents per lb.	Total Value $ million
1968-69	56.0	46.1	25.8	132.8	31.6	42.0	188.8	35.9	67.9
1969-70	62.1	45.1	28.0	131.8	33.9	44.7	193.9	37.5	72.7
1970-71	58.3	54.4	31.7	148.5	39.0	57.9	206.8	43.3	89.6
1971-72	66.3	57.9	38.4	165.1	41.5	68.5	231.4	46.2	106.9
1972-73	76.7	55.5	42.6	177.3	38.0	67.4	254.0	43.3	110.0
1973-74	102.3	57.1	58.4	177.2	36.7	64.9	279.5	44.1	123.4
1974-75	126.1	60.7	76.6	173.0	40.9	70.9	299.1	49.2	147.2
1975-76	142.1	71.9	102.2	167.7	53.0	88.9	309.8	61.7	191.1
1976-77	151.2	82.4	124.6	195.9	66.9	131.0	347.1	73.7	255.7
1977-78	191.0	90.1	172.2	207.6	65.2	135.4	398.7	77.1	307.6
1978-79	229.5	94.9	217.8	224.5	64.2	144.0	454.0	79.7	361.8
1979-80	255.8	95.8	245.2	214.2	57.6	123.4	470.1	78.4	368.6
1980-81	275.0	94.7	260.4	194.5	58.6	114.0	469.6	79.8	374.5
1981-82	379.1	96.8	308.8	198.0	55.7	110.3	517.1	81.0	419.1

Source: SRS, USDA, Mushrooms, August 1982.

The TPSC met in May 1980 to review the request by the Pennsylvania Congressmen for emergency relief and to recommend a response to the President. On the basis of the facts available the TPSC concluded that the industry had not made the case for emergency relief, that relief in this case (in response to political pressure) could set a bad precedent for other industries seeking relief and that the relief options available would be inconsistent with overall U.S. trade policy. Thus the recommendation was to delay any action until the USITC report was available.

TABLE 3
European Community Mushroom Imports, Selected Countries and Total, 1975 - 1980 (Million Pounds Drained Weight Basis)

	1975	1976	1977	1978	1979	1980
China	22.1	17.5	15.6	18.6	30.2	28.6
Korea	2.9	1.9	3.0	9.9	-	4.8
Taiwan	18.7	8.0	10.7	15.0	0.7	2.1
Hong Kong	-	0.2	0.3	1.2	-	0.5
Other	1.0	1.2	1.9	2.7	0.3	1.5
Total	44.7	28.7	31.5	47.4	31.2	37.5

Source: NIMEXE Analytical Tables, Eurostat, 1975-1980 Calendar Year Basis.

No major disparity in views existed among TPSC members at this point. The decision was only to delay a decision until more information, the USITC report, became available. Further, the EC action to liberalize mushroom imports was expected to relieve some of the pressure on U.S. markets. Finally, established policy and precedent argued against action of nature open to the government at this time.

On August 18, 1980 the USITC completed the investigation begun in March and submitted its findings and recommendations to the President. The USITC determined that mushrooms, prepared or preserved (TSUS item 144.20) were being imported into the U.S. in such increasing quantities as to be a substantial cause of injury to the domestic industry. One of the Commissioners did not participate in the findings or recommendations. All four participating Commissioners found serious injury and three recommended relief in the form of quantitative restrictions on imports for a three year period beginning at 86 million pounds on July 1, 1980 and increasing by 9.7 percent in each of the subsequent years. The fourth Commissioner also recommended a quantitative restriction, but for a five year period beginning at 78 million pounds on July 1, 1980 and eventually reaching 104 million pounds in the fifth year. The two recommendations are shown in Table 4.

The USITC found that only the domestic canned mushroom industry produced a product "like" the imported product. They did not address the question of whether fresh mushrooms were "directly competitive" with the imported product. Thus the fresh mushroom industry was excluded from the analysis and hence the findings and recommendations by the Commission. The President had to decide whether to provide relief and, if so, the specific relief measures by October 17, 1980.

TABLE 4
USITC Quota Recommendations, Canned Mushrooms (Million Pounds, Drained Weight Basis)

Quota Period	3 Commis-sioners	1 Commis-sioner
July 1, 1980/June 30, 1981	86.0	78.0
July 1, 1981/June 30, 1982	94.0	78.0
July 1, 1982/June 30, 1983	103.0	86.0
July 1, 1983/June 30, 1984		95.0
July 1, 1984/June 30, 1985		104.0

Source: Estimated

As expected when the USITC report was published with a serious injury finding, the issue again surfaced on the interagency agenda. The trade policy interagency mechanism, which had already been involved in the case, was charged to review the USITC findings, do whatever further analysis and delibertation this body required to be able to arrive at a concrete conclusion as well as, to recommend a decision to the President. The question again before the interagency group was whether relief should be granted and if so what should be its nature. And those whose interests would be affected, pro and con, began to let their concerns be heard.

As the TPSC was moving toward its deadline for making a recommendation, the whole committee was invited to a meeting with Senators Heinz and Schweiker and Congressman Schulze on Capitol Hill. They had arranged a presentation by representatives of the growers and the canners from Pennsylvania. The Congressmen opened the meeting with the argument that

the best way to affirm the U.S. commitment to liberal
trade is to provide the Pennsylvania mushroom growers
relief from imports through a quota. This action in
the face of growing world-wide protectionism would send
the signal that a protectionist game can hurt other
trading partners more than it hurts the U.S.

The industry representatives argued that all they
needed was import relief and some adjustment assistance
over the next five years to allow them to innovate the
available new technology and to make some structural
adjustments. This, they said, would put them again in
a competitive position so that when the quotas were
removed they could compete effectively with the
imported products.

For the most part the TPSC listened, although USTR
responded somewhat sympathetically on the relief
question and Commerce was rather sympathetic on the
adjustment assistance issue. Both, however, indicated
that the TPSC was continuing its deliberations and that
the resolution could not be predicted at that moment.

The TPSC was made aware of the importance of the
issue to the Pennsylvania industry and thus to a
significant portion of the Pennsylvania rural economy
by the actions of the three Pennsylvania Congressmen.
While no direct pressure was exerted either way by the
Administration, the TPSC members were, however, also
made aware that Pennsylvania was an important state and
that a Presidential election was scheduled soon after
the President was required to make a decision on this
case.

As the TPSC continued to gather information no
evidence was found to refute the original findings that
the case for relief was shaky at best on economic
grounds in that the structural change occurring in the
domestic industry was the overriding factor causing the
Pennsylvania industry to be disadvantaged. Import
relief is indicated only when imports are a major cause
of injury and in the absence of all or part of the
imports during an adjustment period, the industry can
recover and thenceforth compete on the basis of former
levels of protection, if any.

Agency concerns about granting relief, apart from
the direct question of whether it would help the
industry, began to solidify. The USDA, recognizing
that the four exporting countries (Taiwan, Korea, Hong
Kong, Peoples Republic of China) represented a combined
export market for U.S. agricultural products of $4-5
billion worried that relief that cut into one of the
few agricutural commodities that these countries could
export to the U.S. could backfire on U.S. exports to
them.

The CEA and the Treasury Department were concerned
with the impact of restricting imports on the level of

inflation. The Trade Act of 1974 includes a provision
(Sec. 202.C.4) that requests the President to take into
account "... the effect of import relief on consumers
and on competition in the domestic markets." The USITC
response to this provision was that consumer costs both
in the form of higher prices and increased consumer
expenditures would be insignificant. Mushroom prices
with relief were expected to average 4.5-8 percent
higher than with no relief. Since mushroom prices in
the absence of relief were expected to increase slower
than the rate of inflation, the additional 4.5-8
percent price increase was expected to have no
significant inflationary impact. Further, with a price
elastic demand, consumers were expected to buy fewer
mushrooms at a higher price. Projections indicated
that the consumer cost would be approximately $191
million with a five year tariff (20 percent-15
percent-15 percent-10 percent-10 percent), $160 million
under the USITC quota recommendation and $142 million
with a three year tariff (20 percent-15 percent-10
percent). At no time during the deliberations did the
TPSC hear directly from consumers or consumer groups on
the issue.

The State Department was reluctant to see any
action taken that would send a negative signal to Korea
where a long standing relationship had become strained
after the assassination of President Park, or to the
PRC where the U.S. was attempting to build a mutually
advantageous relationship. The Department of Labor had
no strong interest in the outcome since a relatively
small number of workers would be affected (the mushroom
industry was not unionized), and they had received only
one letter in support of relief. Concern was also
expressed by USTR, USDA and Commerce that relief action
could entail compensation and that such compensation
might be on sensitive items.

Nevertheless, a subtle change occurred in the way
the issue was being discussed and in the emphasis of
the analysis as the decision date approached. The
discussion took on the tenor that no sound economic
basis seemed to exist to warrant relief, but if the
interagency decision was to be to provide relief, what
should be the nature of that relief? The focus of the
debate centered on quota versus tariff.

The CEA and USDA put together a simple model to
estimate the price elasticity of import demand for
canned mushrooms. This model was used to assess a
tariff rate schedule that would have the same impact on
imports as the quota options being considered. Time
series data used in estimating the model parameters was
for the period 1970/71 through 1978/79. The import
demand functional equation providing the best fit to
the data was (t - Statistics in parenthesis):

$$Q_m = 97.09 - 59.18 \ P_m/P_d + .035 \ Y/PCE$$
$$(\ 1.73) \quad (-3.55) \qquad \qquad (8.62)$$

Durbin-Watson	2.84
R^2	.96
Standard Error of Estimate	7.36

Where:

Q_m = Quantity of mushrooms imported in million pounds (drained weight basis)

P_m/P_d = Ratio of per unit value of imports and domestic per unit value of raw mushrooms

Y/PCE = Disposable income deflated by the personal consumer expenditure index

The derived demand elasticity with respect to price from this equation was -1.24. Thus a one percent increase in the relative import price of canned mushrooms (P_m/P_d) is accompanied by a 1.24 percent decrease in quantity imported. The derived demand elasticity with respect to income was 3.37 thus a 1 percent increase in real disposable income (Y/PCE) is accompanied by a 3.37 percent increase in mushroom imports.

The CEA and USDA using the results of this model assessed the import effect of various combinations of tariff rates. The USITC in their consideration of the case had ruled out the use of tariffs for two main reasons:

1. An across the board tariff could not be used to change export shares of suppliers in desirable ways.
2. The tariff may not be fully passed on by suppliers but may rather be partially absorbed (i.e., the supply curves may not be perfectly elastic).

The first USITC reservation was viewed by the TPSC as a possible advantage since a tariff can be imposed at a level necessary to achieve the desired import reduction without the politically difficult task of assigning import quotas to each supplying country. The second reservation relates to the first in that countries may increase their export market share by absorbing a greater portion of the tariff than other suppliers.

Both Taiwan and Korea recognized these possibilities and both worried about the PRC building market share by absorbing a large portion of the tariff. Both of these countries lobbied U.S. government officials, first not to provide relief, but if relief was to be granted to make it in the form of quotas assigned on the basis of historical market shares. Korea in addition suggested they be rewarded in the quota assignment for having honored their voluntary restraint commitment throughout the immediate historical period.

Table 5 indicates the estimated import levels related to different tariff rate increase options compared to the three year USITC import quota recommendation.

The calculations in Table 5 assume a perfectly elastic supply by exporting countries. To the extent that exporters absorb a part of the tariff increase, import volumes will be higher. The CEA and USDA stated this assumption and its implications during interagency discussions. Not a single agency suggested modification of this assumption or the analysis and the numbers put forward were accepted as definitive. Had there been real opposition to a tariff option this assumption would have been questioned.

TABLE 5
Estimated Mushroom Import Levels Under Various Tariff Rate Options (Million Pounds Drained Weight)

Year July 1- June 30	Projected Imports 1/ No Relief	10% Incr. Tariff 2/	20% Incr. Tariff 2/	20-15-10% Incr. Tariff 2/	ITC Sug- gested Quota
1980-81	115.0	101.0	86.0	86.0	86.0
1981-82	132.0	116.0	99.0	107.0	94.0
1982-83	152.0	133.0	114.0	133.0	103.0

Source: Estimated
1/Assumes a 15 percent annual increase in imports, derived by multiplying the income elasticity of canned mushrooms (3.37) by the latest 3 year average income/PCE index increase of 4.5 percent.
2/Derived by multiplying import demand elasticity by respective tariff rate increases and substracting from projected import levels with no relief.
Note: Tariff increases are in addition to regular duties of 3.2¢/lb plus 10 percent.

The USDA also put forward an alternative to the USITC recommended quota levels in the event that the interagency group were to favor the quota option. USDA suggested quota levels for Taiwan, Korea and "Other" beginning from a baseline of average imports from these areas in the 1978/79-1979/80 period and increasing at the rate of 10 percent per year over a three year period. Table 6 illustrates the USDA quota option.

TABLE 6
USDA Quota Option for Canned Mushrooms (Million Pounds Drained Weight)

	Taiwan	Korea	Other	Total
Base	51.8	26.5	21.7	100.0
1980-81	57.0	29.1	23.9	110.0
1981-82	62.7	32.1	26.2	121.0
1982-83	68.9	35.2	28.9	133.0

Source: Estimated

The advantages of this quota alternative are that (1) it limits the influence of the Taiwan export surge in 1979/80; (2) it provides Korea a market share consistent with recent export levels and provides a small reward for their continuing export restraints; (3) it provides a large enough "Other" category to allow export growth by the PRC. Hong Kong would not receive a quota. Hong Kong does not produce mushrooms but imports them in bulk from Taiwan and the PRC for repackaging and reexport.

Finally the TPSC met to decide upon a recommendation to the President. The TPSC chairman summarized the issue and presented the options of no relief, quotas, or tariff and indicated the need to decide whether or not adjustment assistance would be provided and if so what form it should take. Since the issue was agricultural he asked USDA to be the first to respond.

The USDA position had been formulated carefully to maintain the largest possible measure of flexibility depending on the response of the other agencies. Thus the USDA began by pointing out that the value of relief to the Pennsylvania industry was questionable based on the admittedly rather incomplete economic analysis available. This analysis seemed to indicate that the Pennsylvania industry was being hurt more by increasing domestic competition than by imports. The USDA went on to suggest, however, that other factors must be considered in this decision, including the long history of relief requests stretching back to 1964, all of which had been denied. If the TPSC decided to grant relief then a tariff rather than a quota was felt to be more desirable since the U.S. should attempt to conform as closely as possible to the principles laid down by GATT. A 20-15-10 percent reducing tariff over a three year period appeared to be the most desirable option considering all factors and interests involved. Further, adjustment actions by the industry should be closely monitored to determine whether their adjustment

plans and promises were being fulfilled. This
information would be useful if subsequent relief
requests were made by the industry.

Thus, the USDA position exhibited sensitivity to
a constituent industry concern, focused relief action
on the least damaging option to the other interests
involved, and left the door open to the no relief
option if other agencies felt inclined to argue
strongly for that option. Other agencies concerned
with both the USDA reasoning and position. No agency
argued the no relief option. It was also agreed that
adjustment assistance would be provided, the form and
detail to be determined by a task force composed of
various government agencies that might be able to
contribute.

Announcement of the decision dated October 17,
1980 appeared in the Federal Register of October 23,
1980. The new tariff schedule became effective on
November 1, 1980.

CONCLUSIONS

This case study illustrates several items of
importance to the trade researcher interested in
providing an input into the policy process. First, the
economic analysis is only one of several, sometimes
many, forces bearing on a decision. Legal
requirements, historical precedence, social welfare
considerations and political realities are important,
often overwhelming, factors in the decision process
along with the economic analysis.

Second, while most issues have a long historical
record, points at which decisions must be made arise
quickly, sometimes unpredictably, and the time frame
for analysis is usually extremely short. The mushroom
case analysis began in April 1980 only because of the
political request for emergency relief. Normally, the
total time for work on the case would have been between
the issuance of the USITC report on August 14, 1980 and
the decision deadline of October 17, 1980. Thus the
analysis had to make use of already available data and
information -- there was no time to do primary research
and/or build more than the most simple models. This
is a typical situation and the past work of the trade
researcher will be used if it is readily available,
relevant and in useful form. It helps if the trade
researcher is omniscient!

Finally, the interagency group as a decision body
must be highly operational. It has neither the time
nor the interest in or in many cases the understanding
of methodological expositions or complex arguments.
Information provided must be simple, to the point,
brief and defensible if necessary.

This is not meant to imply that the academic trade researcher has no role in the trade policy decision process of the U.S. Government. It is only to suggest that the role of the trade researcher is at the same time broader than might be realized but also once removed from the heat of the battle.

The role is broader in the sense that agricultural trade policy is no longer, if it ever was, made solely by USDA. Agricultural economists must broaden their base of participation and learn how to interact with and provide useful material for the use of other agencies besides the USDA of importance to the agricultural trade decision process, principally State, Treasury, the CEA and to a somewhat lesser extent Labor and Commerce.

Finally, as has been indicated in the above case study, the time frame for problem-solving decision making is extremely short. The problem-solving economic analysts must draw upon the stock of pertinent subject matter research available at the time that the problem arises. In most cases if the academic trade researcher has contributed to the subject matter stock of knowledge in a way that is useful and relevant to the problem at hand, his work will be used. And he may even be drawn upon to lend first hand knowledge and expertise. This may seem rather haphazard and arbitrary, but in fact the trade researcher, if he is interested in affecting policy decisions, must recognize that he must also play an extension role. Seldom will the results of research be directly and fully applicable to the solution of a given problem. The knowledge gained during the research process, however, can be extremely useful both in terms of a conceptual framework and in terms of the substance in such problem solving analysis.

NOTES

1. This paper was originally completed for
   presentation at the semi-annual meeting of the
   Trade Research Consortium in December 1981. Thus
   The description of the trade policy process
   generally reflects that process as it took place at
   that time. While minor changes have evolved with
   practice, as new issues have arisen and with a
   change in administration, the basic description
   remains accurate. The Mushroom case study
   similarly ends with the announcement by the
   President of the import relief decision of October
   17, 1980. While new developments have taken place
   since that date including a further relief petition
   by the industry, no attempt has been made to
   incorporate this recent history on the basis that
   the new material would not add materially to the
   value of the case study as illustrative of the
   trade policy decision process.

2. Section 22 of the Agricultural Adjustment Act of
   1935, as amended, requires the President to
   establish import quotas on price supported
   commodities, irrespective of existing international
   agreements, whenever imports threaten the ability
   of the government to carry out the domestic price
   support program. Since 1951 the U.S. has had a
   waiver in GATT for the use of Section 22.

3. The Generalized System of Preferences, a provision
   in the Trade Act of 1974, as amended by the Trade
   Agreements Act of 1979, provides for duty free
   entry of imports of certain designated products
   from eligible developing countries. The purpose
   is to help developing countries increase their
   exports, diversify their economies and reduce their
   dependence on foreign aid.

4. Section 201 of the Trade Act of 1974 provides for
   the President to grant temporary import relief
   for the purpose of facilitating orderly adjustment
   to import competition.

APPENDIX

TRADE POLICY STAFF COMMITTEE
Subcommittees and Task Forces
1980

GEOGRAPHIC SUBCOMMITTEES
Western Europe
Japan
Canada and other developed countries
Egypt
Israel
LDCs
Nonmarket economy countries

FUNCTIONAL SUBCOMMITTEES
Antidumping and Subsidies/Countervailing Duty Measures
Standards
Customs and Tariff Matters
Safeguards and GATT Article XIX
Government Procurement Code
Counterfeit Code
Licensing Code/Quantitative Restrictions
GATT Affairs
Aircraft
Agriculture
Fisheries
Steel
Services
Investment
Section 301
Section 337
Export Policy
Generalized Systems of Preferences
Agricultural Commodity Agreements
Industrial Commodity Agreements
Energy Trade
Information Systems
Economic and Trade Policy Analysis
Congressional Liaison
OECD Trade Issues

TASK FORCES
EC Enlargement
Petrochemicals
Winter Vegetables
Annual Report on the Trade Agreements Program
Footwear
Semiconductors
Autos
Color TVs

8
Towards a Countervailing Power Theory of World Wheat Trade

Philip L. Paarlberg and Philip C. Abbott

INTRODUCTION

Empirical studies of the world wheat market have
generally assumed that a neoclassical competitive model
can adequately describe trading behavior (i.e. Rojko,
et al., (1978); Schmitz and Bawden (1973); Takayama
and Judge (1971). Such models assume, at least
implicitly, that no country exercises market power in
trade, even though they are modelling situations where
market power potentially exists. Several researchers
have challenged this assumption by proposing models of
imperfect competition as alternatives: McCalla (1966,
1970); Taplin (1968); Alaouze, Watson and Sturgess
(1978); and Carter and Schmitz (1979).

Evidence on market structure and the existence of
institutions through which market power can be exercised
supports these contentions of imperfect markets.
During the period 1975 to 1979, the United States,
Canada, Australia, Argentina and France supplied an
average of 87.0 percent of world wheat exports, while
47.2 percent of world imports were purchased by the
EEC, Japan, and the Centrally Planned economies. These
concentration ratios are large relative to those found
in industries where oligopolistic market power is
generally assumed to be present, such as the U.S. auto
industry. Institutional evidence on market conduct
indicates that most of these countries control wheat
trade through the use of policy instruments and
institutions capable of exercising market power.
The Wheat Boards in Canada and Australia are clear
examples of institutions which could exercise
market power in trade. The variable levy of the EEC

Philip L. Paarlberg, Trade Policy Branch, International
Economics Division, USDA-ERS and Philip C. Abbott,
Associate Professor, Department of Agricultural
Economics, Purdue University.

and the Japanese resale price are instruments which could be used to capture rents, and state trading in Eastern Europe and many Developing Countries could also be used to exercise market power where market shares are large and competitive assumptions break down.

The progression of literature on the international wheat market suggests that either the extent or the exercise of market power has been changing over the past two decades. McCalla (1966) and Taplin (1969) viewed the international wheat market as a duopoly between the United States and Canada, with Canada acting as a Stackelberg leader. Alaouze, Watson and Sturgess (1978) demonstrated how the stable duopoly of the late 1960's could absorb Australia into a coalition to form a stable triopoly exercising monopoly power. Recently, Carter and Schmitz (1979) have argued that market power lay with the importing countries, particularly with Japan, the EEC and the Eastern European bloc. They reasoned that those regions were acting as oligopsonists by using trade barriers to extract a rent from the international market. A simple market share analysis gives little credence to the hypothesis that market power for exporting countries has vanished, however, and yet the free market policies of the United States and the reduction of stocks held by the major exporters does seem to suggest that those countries have exercised less market power since the middle 1970's.

Several explanations of this progression of events in the world wheat market are plausible and merit investigation. One is that policy instruments including tariffs and quotas are used to achieve domestic policy objectives, such as income redistribution, or to support particular interest groups, and that extraction of rent from the international market is of secondary importance. Hence, governments may be unwilling to change domestic prices to levels necessary to realize "optimal" economic surplus since that would lead to an unacceptable resource allocation (and more importantly, income redistribution) domestically. The policy makers' objective function in that case differs from the economist's standard, (i.e. in Carter and Schmitz) which does not consider income redistribution. A second explanation is that the possibility of retaliation prevents traders from applying even more restrictive policies. The imperfect competition models cited above assume that each country or region trading in the world wheat market accepts its trading partner's policies as a fait accompli. Trade negotiations as have occurred under GATT (General Agreement on Tariffs and Trade) clearly allow for policy retaliation in agricultural commodities. A

third explanation is that the empirical basis for
determining optimal tariffs for the exercise of monopoly
power is weak--both for researchers and for governments.
Hence, governments may be imperfectly implementing
optimal policies due to lack of information.
Furthermore, if the market can be characterized by a
structure as simple as a bilateral monopoly, behaviors
on the basis of which optimal tariffs would be
calculated are indeterminate, and the necessary excess
demand and excess supply functions for countries would
not exist. With several traders possessing potential
market power, this problem becomes even more ambiguous.
Doyle (1981) has presented a duopoly model of the
United States and Canada based on reaction functions
which illustrates the nature and complexity of the
problem, but which is of little use in explaining
recent market conduct and performance.

Several issues emerge which will be the subject
of this paper:

1. What objectives motivate the behavior of the
   major participants in the world wheat market,
   and can an understanding of those objectives
   help in explaining how market conduct has
   evolved over the past two decades?

2. How do domestic and international policies
   interact to jointly determine income
   redistribution to various domestic interest
   groups and the rent extracted from the
   international market?

3. How does a country's behavior in the
   international market depend on its domestic
   policies, policy objectives and the market
   power it may possess? Furthermore, how can a
   policy model of the domestic wheat market be
   integrated with a model of a country's trading
   behavior?

4. If the international wheat market is
   characterized as imperfectly competitive, how
   can the interactions between traders be
   captured in a model?

The purpose of this paper is to explicitly
introduce the income redistribution question into a
model of world wheat trade, where policies are used to
achieve redistributional objectives domestically as well
as to extract rent from the world market. Furthermore,
it will be assumed that larger traders may perceive
and exercise market power in achieving their
objectives. Three sections follow which explore this
question. In the first, a model endogenizing

government policymaking behavior is used to derive
international bargaining functions, which characterize
trading behavior in an imperfectly competitive market,
based upon maximization of a criterion function
exhibiting (potentially) differential treatment of
domestic interest groups and using conjectural
variations theory to capture international market power.
In the second section, parameters of the proposed
government criterion functions and of the conjectural
variations are estimated for the United States, Canada,
and Japan, utilizing existing econometric models. The
third section indicates how these bargaining functions
can be put together to build a model of world wheat
trade.

These results provide a useful approach to
modelling world wheat trade, and strongly suggest that
a game theoretic approach, using conjectural variations,
is preferred. The empirical results also show that
government behavior may not be well explained if one
assumes a government seeks to maximize net social
payoff, and ignores income redistribution.

A POLICY MODEL OF A COUNTRY'S DOMESTIC WHEAT MARKET

In order to explain a country's behavior in the
world wheat market when institutions or policies exist
which can potentially alter or control net trade, a
model of the domestic wheat market treating government
behavior as endogenous is required. Such a model must
consider both domestic and international outcomes, as
policies are intended to affect both. The model
developed in this paper follows the criterion function
approach discussed by Rausser, Lichtenberg and
Lattimore (1983), in that a function representing
government policymakers' objectives is specified as the
weighted sum of welfare accruing to several interest
groups. The government is assumed to set policy at
optimal levels, given its criterion function and
subject to market behaviors. Using the revealed
preference approach of Rausser and Freebairn (1974) the
parameters (weights on interest group welfare) of the
government criterion function can be calculated from
the first order conditions determining optimal policy
and actual market outcomes. The details of our model
specification follow.

Model Specification

We begin with a single-product partial equilibrium
model of the domestic wheat market, incorporating
policy instrument variables in the specification.
Wheat is treated as a homogeneous commodity.
Behavioral equations are:

Production: $\qquad S_i = S_i(P_i^S)$ (1)

Non-Feed Use: $\qquad C_i = C_i(P_i^C)$ (2)

Feed Use: $\qquad F_i = F_i(P_i^F)$ (3)

Private Stock Demand: $I_i = I_i(P_i^I)$ (4)

Public Stocks: $\qquad \Delta G_i = \Delta G_i(P_i^G - P_i^X)$ (5)

Export Demand:$\underline{1/}$ $\qquad X_i^D = X_i^D(P_W)$ (6)

where S_i is the wheat production function on the farm in country i, which is a function of the farmer supply determining price, P_i^S,

C_i is the non-feed consumption of wheat in country i, which is a function of a domestic demand price, P_i^C,

F_i is the feed use of wheat in country i, which is a function of the price faced by feeders, P_i^F,

I_i is the private wheat stock demand in country i, which is a function of the price faced by stockholders, P_i^I,

ΔG_i is the increase in government wheat stocks (supply to government stocks), which is a function of the difference between the price paid for government held stocks, P_i^G, and the market price, P_i^X, and

X_i^D is the net export demand for wheat faced by country i, which is a function of the world market price, P_W. We subsequently call $X_i^D(P_W)$ the international response function, and the form assumed for the function by the government in country i constitutes its conjectural variation.

The following identities link domestic and international prices through policy instruments:

$$P_i^S = P_i^G + \alpha_1^i \qquad (7)$$

$$P_i^C = P_i^X + \alpha_2^i \qquad (8)$$

$$P_i^F = P_i^X + \alpha_3^i \tag{9}$$

$$P_i^I = P_i^X + \alpha_4^i \tag{10}$$

$$P_i^G = P_i^X + \alpha_5^i \tag{11}$$

$$P_i^X = P_i^W + \alpha_6^i \tag{12}$$

where α_1^i is the subsidy to producers above the government stock price (which would be the subsidy above the loan rate in the United States),

α_2^i is the tax on non-feed use (flour milling),

α_3^i is the tax on feed use,

α_4^i is the tax on commercial stock acquisitions,

α_5^i is the difference between the government stock price (U.S. loan rate) and market price, and

α_6^i is the export subsidy.

These identities may be rewritten in terms of the world market price, P_W, as follows:

$$P_i^S = P_W + \alpha_6^i + \alpha_1^i + \alpha_5^i \tag{13}$$

$$P_i^C = P_W + \alpha_6^i + \alpha_2^i \tag{14}$$

$$P_i^F = P_W + \alpha_6^i + \alpha_3^i \tag{15}$$

$$P_i^I = P_W + \alpha_6^i + \alpha_4^i \tag{16}$$

$$P_i^G = P_W + \alpha_6^i + \alpha_5^i \tag{17}$$

In addition to these, a market clearing identity or equilibrium condition relating supplies and demands must also be specified. Hence:

$$X_i^S = X_i^D = S_i + (I_i^{t-1} - I_i^t) - C_i - F_i - \Delta G_i \tag{18}$$

where X_i^S is net export supply of country i.

Government policy determines economic surplus to each actor (interest group) as well as revenue to the government, given these behavioral relationships and identities. The assumed government criterion function is therefore a weighted sum of the welfare (operationally, economic surplus) accruing to each interest group (producers, flour milling/consumers, feeders/consumers, stockholders, and the government itself). Any rents in the international marketplace will be captured as economic surplus to various interest groups or as tariff revenue to the government. Our criterion function, therefore, is:

$$W_i = \gamma_i^P \text{ (Producer Surplus}_i) + \gamma_i^C \text{ (Flour Milling/}$$
$$\text{Consumer Surplus}_i)$$
$$+ \gamma_i^F \text{ (Feeder/Consumer Surplus}_i) + \gamma_i^I \text{ (Stock-}$$
$$\text{holder Surplus}_i)$$
$$+ \gamma_i^G \text{ (Net Government Revenue}_i) \qquad (19)$$

where surplus is the appropriate area under a demand curve or above a supply curve and γ_i^X is the weight the government assigns to this welfare measure for group x in country i. Hence

$$W_i = \gamma_i^P \int_0^{P^*} S_i(P_i^S) dP_i^S - \gamma_i^C \int_0^{P^*} C_i(P_i^C) dP_i^C$$

$$- \gamma_i^F \int_0^{P^*} F_i(P_i^F) dP_i^F - \gamma_i^I \int_0^{P^*} I_i(P_i^I) dP_i^I +$$

$$\gamma_i^G [\alpha_{2i} C_i + \alpha_{3i} F_i + \alpha_{4i} I_i - (\alpha_{1i} + \alpha_{5i}) S_i$$

$$- \alpha_{6i} x_i^D] \qquad (19a)$$

Optimal Policy

The problem to be solved by a government is to find a set of optimal policies which maximize welfare W_i subject to the six behavioral equations, (1) - (6), the market clearing condition (18), the identities linking prices, (13) - (17), and the definitions of our surplus measures. Since one behavioral equation is the net export demand function, or international response function $X_i^D(P_W)$ this country is not assumed to behave competitively in the world market. Rather, the country sees (or makes a conjecture on) the response of its trading partners to its policies and adjusts its policies accordingly, as a monopolist or oligopolist would (Bresnahan, 1981). If the country were a net importer, the net export demand function would be interpreted as a net import supply function, as net exports would be negative. In principle the conjecture $X_i^P(P_W)$ depends on policies or reactions abroad.

Assuming standard definitions of economic surplus, first order conditions for optimal policy selection are:

$$\frac{\partial L_i}{\partial \alpha_1^i} = \gamma_i^P S_i - \gamma_i^G (\alpha_1^i \frac{\partial S_i}{\partial P_i^S} + S_i) + \lambda_i (\frac{\partial S_i}{\partial P_i^S}) = 0 \quad (20)$$

$$\frac{\partial L}{\partial \alpha_2^i} = -\gamma_i^C C_i + \gamma_i^G (\alpha_2^i \frac{\partial C_i}{\partial P_i^C} + C_i) -$$

$$\lambda_i (\frac{\partial C_i}{\partial P_i^C}) = 0 \quad (21)$$

$$\frac{\partial L}{\partial \alpha_3^i} = -\gamma_i^F F_i + \gamma_i^G (\alpha_3^i \frac{\partial F_i}{\partial P_i^F} + F_i) -$$

$$\gamma_i (\frac{\partial F_i}{\partial P_i^F}) = 0 \quad (22)$$

$$\frac{\partial L}{\partial \alpha_4^i} = -\gamma_i^I I_i + \gamma_i^G (\alpha_4^i \frac{\partial L_i}{\partial P_i^I} + I_i) -$$

$$\lambda_i (\frac{\partial I_i}{\partial P_i^I}) = 0 \quad (23)$$

$$\frac{\partial L}{\partial \alpha_5^i} = \gamma_i^P S_i - \gamma_i^G (\alpha_5^i \frac{\partial S_i}{\partial P_i^S} + S_i) +$$

$$\lambda_i (\frac{\partial S_i}{\partial P_i^S} - \frac{\partial G_i}{\partial \alpha_5^i}) = 0 \quad (24)$$

$$\frac{\partial L}{\partial \alpha_6^i} = \gamma_i^P S_i - \gamma_i^C C_i - \gamma_i^F F_i - \gamma_i^I I_i +$$

$$\gamma_i^G (\frac{\partial C_i}{\partial P_i^C} \alpha_2^i + \frac{\partial F_i}{\partial P_i^F} \alpha_3^i +$$

$$\frac{\partial I_i}{\partial P_i^I} \alpha_4^i - (\alpha_1^i + \alpha_5^i) \frac{\partial S_i}{\partial P_i^S} - X_i^D) +$$

$$\lambda_i (\frac{\partial S_i}{\partial P_i^S} - \frac{\partial C_i}{\partial P_i^C} - \frac{\partial F_i}{\partial P_i^F} - \frac{\partial I_i}{\partial P_i^I}) = 0 \qquad (25)$$

$$\frac{\partial L}{\partial P_W} = \gamma_i^P S_i - \gamma_i^C C_i - \gamma_i^I I_i + \gamma_i^G (\alpha_2^i \frac{\partial C}{\partial P_i^C} +$$

$$\alpha_3^i \frac{\partial F_i}{\partial P_i^F} + \alpha_4^i \frac{\partial I_i}{\partial P_i^I} - (\alpha_1^i + \alpha_5^i) \frac{\partial S_i}{\partial P_i^S} -$$

$$\alpha_6^i \frac{\partial X_i^D}{\partial P_W}) + \lambda_i (\frac{\partial S_i}{\partial P_i^S} - \frac{\partial C_i}{\partial P_i^C} - \frac{\partial F_i}{\partial P_i^F} - \frac{\partial I_i}{\partial P_i^I} -$$

$$\frac{\partial X_i^D}{\partial P_W}) = 0 \qquad (26)$$

$$\frac{\partial L}{\partial \lambda_i} = S_i + I_i^{t-1} - C_i - F_i - I_i - C_i - X_i^D = 0 \quad (27)$$

where λ_i is the Lagrange multiplier on the market
clearing condition equation (18), L is the Lagrangian
function for this problem, and other variables are as
previously defined.

When the weights on interest group welfare γ_i^j are
given, these eight equations may be simultaneously
solved for optimal levels of the policy variables, α_j^{i*},
as well as for the world price, P_W, and the Lagrange
multiplier, λ_i. Given the solution for these variables,
the remaining variables can be found using (1)-(18).
If one is willing to assume that this model describes
the constraints faced by government policymakers and

that the government is indeed optimizing, that solution
may be used to forecast government behavior, and hence
domestic market behavior in response to changing
international market conditions. This solution will
capture any market power, since these first order
conditions were derived given an assumed (or perceived)
export market behavior through inclusion of equation (6)
and (18). These conditions and assumptions can also be
used to estimate the revealed preference by government
policymakers for welfare accruing to each interest
group. In addition, indirect tests on whether or not a
country behaved as an oligopolist or oligopsonist can
be conducted by applying these first order conditions to
actual market outcomes. Using the derived weights,
assumptions on international market responses consistent
with observed behaviors can be determined.

International "Bargaining" Functions

Optimal policies can be expressed as a function of
the world market price P_W and the expected response
function in the international market, $X_i^D(P_W)$. These
may be substituted into the market behavioral equations
(10)-(5) and the market clearing identity may be used,
via the implicit function theorem, to derive an export
supply function X_i^S for a country which endogenizes the
government's behavior and has as its arguments the
world market price P_W and the derivative of the
international response function $\frac{\partial X_i^D(P_W)}{\partial P_W}$ which is the
conjectural variation of the government in
country i. This function is given by:

$$X_i^S = F\left[P^W,\ \gamma_i^P,\ \gamma_i^C,\ \gamma_i^F,\ \gamma_i^I,\ \gamma_i^G,\ \frac{\partial S_i}{\partial P_i^S},\ \frac{\partial C_i}{\partial P_i^C},\ \frac{\partial F_i}{\partial P_i^I},\right.$$

$$\left.\frac{\partial I_i}{\partial P_i^I},\ \frac{\partial G_i}{\partial P_i^G},\ \frac{\partial X_i^D}{\partial P_W}\right] \tag{28}$$

Since the partial derivatives $\frac{\partial S_i}{\partial P_i^S}, \frac{\partial C_i}{\partial P_i^C}, \frac{\partial F_i}{\partial P_i^F}, \frac{\partial I_i}{\partial P_i^I}, \frac{\partial G_i}{\partial P_i^G}$
are characteristic of domestic market behavior, and if
the weights γ_j^i are treated as known parameters, then
this function may be simplified to:

$$X_i^S = F\left[P^W,\ \frac{\partial X_i^D}{\partial P_W}\right] \tag{29}$$

If but one country is setting policy based upon

responses of its trading partners, then the excess demand and supply functions are uniquely determined and so is $\frac{\partial X_i^D}{\partial P_W}$. This would be the case of pure monopoly or pure monopsony. With this parameter uniquely determined, X_i^S, W_i and P_W can be found by solving equation (29) and the international response function (equation (6)) simultaneously. However, if more than one country is setting policy based upon responses by trading partners the excess demand and excess supply functions are not uniquely determined. In this instance $\frac{\partial X_i^D}{\partial P_W}$ is ambiguous and the solution to the first-order condition is indeterminate. Hence, a game theoretic framework in which the optimal policy choice depends upon policy decisions elsewhere is required. One such approach is to assume that each government with market power guesses the reactions of its trading partners, so that the assumed response is a conjecture on the variation in net export demand. This is the approach taken in conjectural variation theory (Bresnahan, 1981). Other game theoretic approaches could also be employed to resolve the indeterminary of $\frac{\partial X_i^D}{\partial P_W}$ as well.

OBSERVED GOVERNMENT POLICY BEHAVIOR IN THE WHEAT MARKET: THE UNITED STATES, CANADA, JAPAN

This section presents and discusses the implications of the weights on the different economic surpluses which comprise the calculated government criterion function. Using existing econometric models (McKinzie (1979), Jabara (1981), and Baumes and Meyers (1980)), and observed behavior, the weights for the United States, Canada, and Japan are calculated. These weights are used to test hypotheses on stability and pricing behavior in world markets.

Interest Group Weight Calculation and Interpretation

Before analyzing the calculated values of the weights, their interpretation needs to be clarified. The weight on producer surplus is the partial derivative of the criterion function with respect to producer surplus. Hence, γ_i^P can be interpreted as the marginal value the policy-maker attributes to producer welfare.[2/] In a similar manner, the remaining weights may be interpreted as the marginal value the policymaker places on the welfare accruing to each particular interest group.

The weights can be calculated from the first order conditions using revealed preference theory (i.e. Rausser and Freebairn (1974)). Assuming the policy-maker is rational, consistent, and faces certainty, if the criterion function accurately represents the

objective of the policymaker, then the observed
behavior represents an equilibrium which satisfied the
first order conditions. Econometric analysis of the
United States(Baumes and Meyers), Canada (McKinzie),
and Japan (Jabara) provides estimates of the partial
derivatives, while the policy choices and quantities
are given by observable data. Substituting this
information into the first-order conditions, and
setting the producer weight equal to unity yields a
system of equations which can be solved for the
remaining relative weights, γ_i^X/γ_i^P. The producer weight
is selected as the numeraire because ostensibly
policies in the three countries considered are producer
oriented. In principle, the relative solution values
are indifferent to which weight is chosen as numeraire.
Problems would arise however, if the numeraire weight
were zero.

U.S. Weights

The marginal values of various interest groups'
welfare relative to producers in the United States
from 1960 to 1978 are shown in Table 1. The general
pattern for these relative weights is that they have
declined during the 1960's, then rose in the middle
1970's, and finally declined slightly in the late
1970's. Plausible relationships between the calculated
weights and policy changes exist and are presented
below. Although the weights are calculated from the
first order conditions, it is crucial to remember the
direction of causality. The policies implemented are
a function of the values of the weights, which are a
result of the political-economic environment. The
political factors which prompted changes in the values
of the weights are not discussed in this paper. These
political influences need to be determined if this
model is to be used as a predictive tool.

The relative weight on non-feed use, which
represents the consumer surplus on direct human
consumption, varied between a high of 1.26 in 1973 to
a low of .89 in 1968. At the beginning of the data
period, the marginal values of non-feed use and
producer welfare were approximately equal, hence, a
value of the relative weight on non-feed use of about
one. In 1964, the relative value of non-feed use
declined sharply. Also,the value the U.S. government
placed on its own welfare declined. In that same year
a major shift in U.S. policy occurred. Beginning in
1964, a tax was imposed on flour processors in the
United States. Further, the United States implemented
the export and domestic marketing certificate system
which substituted direct U.S. government assistance to
farmers for the indirect assistance via the loan rate

TABLE 1
United States--Marginal Value of Various Interest
Groups Relative to Producer (γ^X/γ^P)

Year	Non-Feed Use	Feed Use	Private Stock	Government
1960	1.0070	1.1630	1.0657	1.0038
1961	.9945	.8510	.9470	.9969
1962	1.0097	.9477	.9869	1.0072
1963	1.0162	3.1460	1.2177	.9948
1964	.9190	1.2752	1.2189	.9793
1965	.9341	1.2166	1.2311	.9847
1966	.9706	1.4155	1.2687	.9950
1967	.9499	1.7771	1.1340	.9862
1968	.8895	.9366	.9448	.9538
1969	.8949	.9661	.9606	.9491
1970	.9035	.9587	.9570	.9546
1971	.9476	1.0818	1.0878	.9839
1972	1.0495	1.3146	1.2754	1.0250
1973	1.2616	2.4293	2.3638	1.0365
1974	1.2372	4.1478	1.7482	1.0379
1975	1.1429	3.4966	1.3407	1.0142
1976	1.0424	1.2882	1.0580	1.0109
1977	1.0112	1.0694	1.0222	.9863
1978	1.0460	1.2111	1.0933	.9937

Source: Calculated using econometric model of Baumes
and Meyers (1980).

system. By 1968 and 1969, the relative value of non-
feed use to the U.S. policymaker had declined from
unity to .89. Similarly, the relative value of U.S.
government revenues declined into the late 1960's.
Although the export marketing certificate was
discontinued in 1966, the domestic marketing
certificate was linked to 100 percent of parity, the
domestic flour processing tax and export subsidies were
increased, and large government stock purchases were
made.
     Beginning in the early 1970's, the weight of non-
feed use begins to rise as does the weight on U.S.
government revenues. Thus, the U.S. government appears
increasingly reluctant to finance direct income
transfers to U.S. growers in response to the exogenous
shocks to the system during this period. As a result,
the flour processor tax gradually declined. In 1968
the tax was $42.46 per ton. However, by 1972 the tax

had been reduced to $35.46 per ton. Further, the U.S. government reduced the level of support to producers from about $46 per ton in 1969 to $19 per ton in 1971. Also, the government adopted a policy of reducing the burdensome stocks accumulated in 1969 and 1970.

The highest relative weights on non-feed use were in 1973 and 1974 at 1.26 and 1.24, respectively. In 1973, the export subsidy was eliminated, and in 1974 the flour processing tax was eliminated. Not surprisingly, the weight on U.S. government welfare is also the highest in these two years. The U.S. government did not subsidize producers during this period. Following 1974, the weights on non-feed use and government welfare begin to decline, and the target price-deficiency payment policy was implemented. By 1978, the weight on non-feed use is slightly higher than in 1960, as is the weight on U.S. government welfare. In contrast to the loan rate program of the 1960's, the target price system does not involve indirect income transfers from consumers to producers and does not result in large public stock purchases. The above discussion of the weight on non-feed use noted shifts in the U.S. government's evaluation of its own welfare, hence, a separate discussion of that weight will not be included. However, there are two other weights calculated for the United States which require a brief discussion--the relative value of feeders/consumers, and private stockholders. The temporal patterns in these two weights are similar to non-feed use, with the high point of both in the 1972-74 period. This reflects the inflationary meat price situation during the second Nixon administration, and the low level of world grain stocks in those years.

Canadian Weights

The relative marginal values of various interest groups in Canada are shown in Table 2. The common pattern of these weights is that they have been rising over time, and all are below the levels calculated for the United States. Both results are generally to be expected. The Canadian Wheat Board (CWB) is a producer oriented board, hence, a priori producers would be expected to be given a relatively higher weight by the Board than in the United States. Further, the CWB is not completely independent of Federal policy and consequently pressures put on the Federal Government in the 1970's should be revealed as an increasing consumer weight.

The relative weight on non-feed use varies between a low of .73 in 1972 and a high of .99 in 1975. This weight drops suddenly in 1969, and an implicit tax on flour millers was imposed in that year. The relative weight on non-feed use was the lowest in 1972--the year

TABLE 2
Canada--Marginal Value of Various Interest Groups
Relative to Producers (γ^X/γ^P)

Year	Non-Feed Use	Feed Use	Private Stock	Government
1960	.8151	.8665	.9296	.9478
1961	.7657	.7089	.9142	.8935
1962	.8279	.8652	.9178	.9590
1963	.8496	.8893	.9451	.9675
1964	.8404	.8696	.9353	.9604
1965	.8277	.8518	.9221	.9555
1966	.8487	.8846	.9524	.9677
1967	.8664	.8733	.9454	.9574
1968	.8073	.8413	.9322	.9435
1969	.7676	.8976	.9516	.9576
1970	.7900	.8776	.9234	.9300
1971	.8144	.9108	.9507	.9588
1972	.7278	.8099	.8984	.9649
1973	.9428	.9132	.9364	.9655
1974	.9599	.9158	.9374	.9677
1975	.9940	.9600	.9670	.9853
1976	.9718	.9893	.9953	.9966
1977	.9113	.9677	.9822	.9876

Source: Calculated using econometric model of McKinzie
(1979).

the implicit tax was the highest.   This weight increases
from .72 in 1972 to .94 in 1973.   From 1973 to 1977
the mill price was below the export price, thus
implying a subsidy to flour millers.
     The relative weights on feeders/consumers is
roughly stable in the middle to upper .80's until 1971.
Thereafter, it begins to rise.   Correspondingly, the
implicit feeding subsidy of off-board pricing was
fairly stable until 1971 (in real terms).
     The values for the weights in 1970 suggest the
shift in policy directions undertaken by the LIFT
program (Lower Inventories for Tomorrow), but the
moderate changes do not suggest the radical changes
which resulted in a halving of Canadian wheat area.
The relative value of on-farm stocks is lower.   The
relative value of feed use is also lower as the
implicit subsidy to feeders declined.   The value of
non-feed use is higher, and the implicit tax lower. The
value the Board puts on its welfare was .96 in 1969 and

.93 in 1970.

Japanese Weights

The relative weights on consumers and the government for Japan are given by Table 3. These data indicate that relative to producers, Japanese consumers fare poorly. Comparison to Canada and the United States reveal the extent to which Japanese policy favors producers. Over the past two decades the consumer weight started at about .8 and has steadily declined to .4, with the exception of 1973 when world prices rose above resale prices, thus giving a temporary consumer subsidy. The relative weight the

TABLE 3
Japan--Marginal Value of Consumer and Government Welfare Relative to Producers (γ^X/γ^P)

Year	Consumers	Government
1960	.7928	.9265
1961	.7986	.9335
1962	.8353	.9243
1963	.8065	.8456
1964	.8364	.8994
1965	.8166	.8897
1966	.8140	.8611
1967	.7949	.8505
1968	.7966	.8386
1969	.7506	.7893
1970	.6899	.7038
1971	.6447	.6606
1972	.5521	.5600
1973	.7108	.5877
1974	.5261	.4781
1975	.4785	.4419
1976	.3982	.4104

Source: Calculated using econometric model of Jabara (1981).

Japanese government puts on its welfare has also declined steadily over the past twenty years. In 1960, the government weight was about .93, but by 1976 had declined to .41. These data confirm the apparent willingness of the Japanese government to support the small number of producers through large income transfers from the Treasury.

Equality and Stability of the Calculated Weights

When doing modelling of the world wheat market, researchers usually implicitly assume the government attempts to maximize net social payoff. Thus, the values on the weights for consumers, feeders and stockholders, and the government are stable and equal to unity. The calculated weights in Tables 1 to 3 suggest these assumptions are not valid and the usefulness of net social payoff as the government criterion function is limited. To formally evaluate this hypothesis the stability and equality of the mean value of the weights are tested.

Table 4 reports the results of the stability tests for all the weights in the three countries. The data is divided into two sets, 1960-1971 and post-1971. The hypothesis is that the means of the two data periods are equal, while the alternative is that they are not. In the United States with exception of the relative weight on feed use, the hypothesis of stability can be rejected. In Canada, the hypothesis can be rejected on all but the on-farm stocks weight, while in Japan the stability of means hypothesis can be rejected in both instances. Thus, these tests suggest that only the weights for feed use in the United States and on-farm stocks in Canada can be assumed to be stable. Hence, it would seem that government policy objectives have been changing over time and any government criterion function needs to incorporate these changes.

Table 5 reports the tests for the mean of the calculated series equal to one, which is equivalent to testing for equality of the weights, which is the standard criterion function assumed in the optimal tariff calculation. Equality is required if one is to assume income redistributional objectives are not an important part of the government criterion function. For the United States the hypothesis that the mean equals one cannot be rejected for only non-feed use and the government weight. In Canada and Japan, the hypothesis can be rejected in all instances. Hence, one must conclude that the assumption that the weights on consumers, government, and stockholders equal one is in general not reasonable for these countries.

Indirect Tests of Perceived Market Power

Given the values for the weights in the criterion function, the first order conditions can be used to solve for the trade elasticities perceived by decision-makers when setting policy. These elasticities are shown in Table 6. For Canada the elasticity is quite variable, ranging from 1.0 to -6.2, with a mean value

TABLE 4
Stability Test on Relative Marginal Values: United States, Canada, Japan

Group	United States		Canada		Japan	
	t	Critical value*	t	Critical value*	t	Critical value*
Non-Feed	-4.734	-2.11	-3.292	-2.12	5.993	2.131
Feed	-1.928	-2.11	-2.311	-2.12	--	--
Private Stocks	-2.262	-2.11	-1.547	-2.12	--	--
Government	-3.499	-2.11	-2.996	-2.12	7.772	2.131

* at a .025% significance level.

TABLE 5
Test for Mean of Relative Marginal Values Equal to Unitary, United States, Canada, Japan

Group	United States		Canada		Japan	
	t	Critical value*	t	Critical value*	t	Critical value*
Non-Feed	.498	2.101	8.295	2.110	-8.660	-2.12
Feed	2.777	2.101	7.936	2.110	--	--
Private stocks	2.650	2.101	10.491	2.110	--	--
Government	-.972	2.101	7.626	2.110	-5.843	-2.12

*at a .025% significance level.

TABLE 6
Implied Trade Elasticities Faced by   Canada, Japan and
United States

Year	Canada	Japan	United States
1960	-1.054	9.255	2.355
1961	-.357	8.020	1.208
1962	-2.654	9.682	1.424
1963	-1.951	7.140	1.208
1964	-2.065	8.076	2.987
1965	-1.382	6.052	.897
1966	-2.189	5.496	.990
1967	-1.510	5.845	1.440
1968	-.439	4.214	40.454
1969	.275	3.671	-.559
1970	1.040	3.255	.807
1971	-.098	2.592	.919
1972	-.515	2.318	.558
1973	-1.832	4.849	1.322
1974	-1.651	5.370	1.736
1975	-3.101	2.688	1.766
1976	-6.194	2.622	2.111
1977	-1.460	-	.416
1978	-	-	.429
Mean ($\overline{X}$)	-1.508	5.364	3.288
Std. Dev. of $\overline{X}$	.371	.583	2.073

of -1.508. The variability in the annual trade
elasticity implies that either Canadian policymakers
treat the elasticity of excess demand they face as an
unknown or that reactions by competitors are variable
over time. Thus, a game theoretic framework which
views the elasticity as an endogenous variable appears
to be more appropriate for Canada than the neoclassical
approach.

The excess supply elasticity perceived by Japan
has a mean value of 5.4, and appears to be generally
declining over the past two decades. Between 1960 and
1964 this elasticity ranged from 4.7 to 7.1. From
1965 to 1968, it ranged between 6.1 and 4.2. Thereafter, the excess supply elasticity has been perceived
by Japan to lie between 3.7 and 2.3, except for the
years 1973 and 1974. The decline in perceived
elasticity by Japan appears to be correlated with the

rising market power of Japan in world wheat trade. This result is clearly consistent with the notion that Japan behaved more or less as a competitor in the earlier 1960's and now perceives that it has monopsony power. Carter and Schmitz (1979) argue Japan has set tariff levels consistent with the optimum level. Using the traditional optimal tariff calculation, these data imply an optimum import tariff level of roughly 40 percent in recent years. Data for 1976 show the resale price to consumers is roughly 58.62 percent above import prices. These weights suggest that the Japanese policy is also a result of domestic considerations.

The elasticity perceived by the United States as calculated from the first order conditions has a mean of positive 3.288, and is positive except for 1969. This sign is contrary to prior theoretical expectations. Testing the hypothesis that the elasticity is equal to zero, one finds that the hypothesis cannot be rejected. Apparently, U.S. policy is set solely for domestic purposes with the view that the demand for U.S. wheat exports is perfectly inelastic. These results are consistent with the attitude expressed by many former and current USDA officials that the United States is a "residual" supplier and will export the difference between foreign demand and foreign supplies. (See USDA, "Analysis of the American Agricultural Movement's Parity Pricing Proposal," 1978) It is also consistent with the policy of the high price floor established for U.S. commodities during the 1950's and 1960's. The export subsidy program begun in the 1950's was implemented only after the United States began to realize the loss in exports due to high price floors and the high costs of burdensome public stocks. Thus, it appears that U.S. policy behavior can be analyzed assuming the perceived slope is zero.

EQUILIBRIUM IN A GAME THEORETIC FRAMEWORK

The first section of this paper illustrated the derivation of international bargaining functions from the first order conditions for optimum policy. These functions incorporate the countries' strategy choice set, and include the term $\frac{\partial X_i^D}{\partial P_W}$ --the international response function slope or the conjectural variation of the government in country i. If more than one country sets optimal policy recognizing the actions of trading partners, then this term is not uniquely determined and a game theoretic framework is appropriate.

Given the policy choice sets, market outcomes may be found for all countries by defining appropriate equilibrium conditions. The first equilibrium condition is straightforward, and applies for all alternatives. World exports must equal world imports:

$$\sum_{i=1}^{n} X_i^S = 0 \tag{30}$$

One type of equilibrium solution can be found by assuming that each country sets its policies with the view that no other country will retaliate or that previous period data may be used to evaluate international responses. Hence, reaction functions of the form:

$$X_i^S = F\ (P_W,\ 0) \quad \text{or} \quad X_i^S = F\ (P_W,\ \left[\frac{X_i^D}{P_W}\right]_{t-1}) \tag{31}$$

may be assumed. [3/] Equation (31) is then utilized to solve for the world market price. This approach is similar to the reaction function approach proposed by Doyle. This is analogous to the Cournot-Nash solution.

An alternative to the Nash equilibrium is the Stackelberg equilibrium which assumes one country accepts a leadership role by selecting an optimal set of policies subject to the other countries' reaction functions. In this case, the leader would examine its welfare $W_i\ (P_W,\ \frac{\partial X_i^D}{\partial P_W})$ subject to equation (31) and the reaction functions for all other countries as previously defined. A common characteristic of the Stackelberg solution is its tendency to degenerate into trade warfare when the leader-follower roles are no longer acceptable. The oligopoly models of the wheat market proposed by McCalla (1966), Taplin (1969), and Alauoze, Watson, and Sturgess (1978) were similar to this type of solution alternative. McCalla (1970), and Alaouze, Watson, and Sturgess discuss trade warfare in the context of their models.

Neither the Nash nor the Stackelberg solutions discussed above allow for the possibility of coalition formation. The Stackelberg solution may be modified to incorporate possible coalition(s), however. Given the prior concerns on income distribution, however, one must now also be concerned that coalitions do not give equal weight to all members of the coalition(s). Hence, the coalition might act to maximize its welfare such that:

$$W^S = \sum_{i\ \text{in}\ S} \eta_i W_i, \tag{32}$$

where η_i are the subjective values on the welfare accuring to members of the coalition, subject to equation (32) and the reaction functions of non-coalition members. Alternatively, two coalitions may form and maximization of coalition welfare for each could yield reaction functions to be solved in Nash fashion as previously described.

Conjectural variations theory presents an alternative to these older and more standard game theoretic approaches. In that case, a guess, or conjecture on the part of government is used to define $\frac{\partial x_i^P}{\partial P_W}$, and an equilibrium may be determined using the bargaining functions for each country with $\frac{\partial x_i^D}{\partial P_W}$ set and using equation (32) to close the model. A set of consistent conjectures could be found using the derived bargaining functions for other countries to derive the actual response on the part of other traders. That response would be found by simply differentiating the bargaining functions with respect to the world price and summing responses. The Nash and Stackelberg models presented above posit inconsistent conjectures between actual responses and assumed responses on the part of governments with market power. They nevertheless fit neatly into the conjectural variations framework. Other models of inconsistent conjectures are also possible, and might be found by, for example, regressing the estimated conjectures of Table 6 on policy variables of the trading partners. Coalition formation may be captured by assuming joint determination of conjectures and joint maximization by colluding trading partners.

CONCLUSIONS

This paper has developed a model of the world wheat market endogenizing government policy formation. The model relies upon government criterion functions to yield a set of first order conditions which determine a country's policy choice set. From these two functions, representing a country's international market behavior and welfare, market outcomes are derived utilizing alternative market equilibrium conditions. Equilibrium conditions can be standard competitive or monopolistic market assumptions or game theoretic conditions. This model allows for simultaneous endogenous policy formation in all countries, policy retaliation, and for coalition structures to evolve and dissolve. Hence, it has explicitly incorporated government behaviors intended to achieve domestic objectives given imperfect international markets.

Analysis of Canada, Japan and the United States suggests that the values government policymakers place on the welfare of various interest groups are not equal, and have changed over time supporting the notion that income redistributional objectives are important and policy objectives are not static. Hence, maximization of net social payoff (as is commonly assumed) is an inappropriate government criterion--income redistribution

needs to be explicitly considered. In the United
States, the weights at the beginning of the 1960's are
close to unity for all interest groups. Throughout the
decade of the 1960's the values decline. With the
inflationary problems of the early and middle 1970's,
the value the U.S. policymaker places on consumers,
livestock feeders, and taxpayers rises above unity.
Although relative to producers, other Canadian
political interest groups fare less well than their U.S.
counterparts, the changes over time are similar. For
Japan, wheat policy entails substantial income
transfers from consumers and taxpayers to producers,
and the pattern of weights reflects this policy.
Throughout the 1960's and 1970's, the relative weights
on these two interest groups have fallen from 0.79 and
0.93 in 1960, respectively, to 0.40 and 0.41 in 1970.
Further, given the calculated weights, the perceived
trade elasticity (conjectural variations) faced by an
exporter or importer can be determined. The trade
elasticity estimates for Canada imply that country
views the world wheat market as a game in determining
policy. In contrast, Japan does not appear to
formulate policy with the trade elasticity as variable,
although her perceived market power seems to be
increasing. It appears that the United States develops
policy in the wheat market with the assumption that the
excess demand it faces is perfectly inelastic.

Future research is needed to test the weights on
the government criterion function for other major
wheat trading nations, such as Australia, Argentina,
and the EC, and to determine the perceived trade
elasticities for these nations. Depending upon the
outcomes of these tests, econometric models of the
wheat sectors in some countries may have to be
redesigned. Finally, the world wheat market needs to
be solved in a game theoretic framework and possible
coalition structures need to be examined in order to
capture the potential reactions in international
markets to domestic policy changes in forecasts of
wheat market behavior.

NOTES

1. $X_i^D(P_W)$ is the perceived international market
behavior faced by country i. By definition, X_i^D must
equal the export supply of country i, we shall call X_i^S.
That supply is derivable from domestic behavioral
relationships and a market clearing equilibrium
condition, given policy.

2. Obviously, the government criterion function
may be altered by a monotonic transformation leaving
optimal policy and market outcomes unchanged.
Therefore, we will subsequently consider relative

weights (γ^X_i/γ^P_i) since the absolute values of these weights are meaningless.

3. It is worth reminding the reader that $x^S_i = x^D_i$ by definition. Hence, equation (32) may be used to determine $\frac{\partial x^D_i}{\partial P_W}$ by remembering

$$x^D_i = \sum_{j=i} x^S_j$$

REFERENCES

Alaouze, C.M., A.S. Watson, and N.H. Sturgess. "Oligopoly Pricing in the World Wheat Market." American Journal of Agricultural Economics. 60 (1978): 173-185.

Baumes, H.S. and W.H. Meyers. "The Crops Model: Structural Equations, Definitions, and Selected Impact Multipliers." NED Staff Report. ESCS/USDA. March 1980.

Bresnahan, T.F. "Duopoly Models with Consistent Conjectures." American Economic Review. 71 (1981): 934-945.

Carter, C. and A. Schmitz. "Import Tariffs and Price Formation in the World Wheat Market." American Journal of Agricultural Economics. 61 (1979): 517-522.

Doyle, J. "A Study of Stability Conditions for the Imperfect World Wheat Market: A Duopoly Simulation Model." Journal of Rural Development. 4 (1981): 37-53.

Grennes, T., P.R. Johnson, and M. Thursby. The Economics of World Grain Trade. New York: Praeger Publishers, 1978.

Jabara, C.L. "Interaction of Japanese Rice and Wheat Policy and Impact on Trade." IED Staff Report, ESS/USDA. April 1981.

Japan Flour Millers Association. "Japan Wheat Import and Pricing Policies." FDCD Working Paper, ERS/USDA. May 1978.

McCalla, A.F. "A Duopoly Model of World Wheat Pricing." Journal of Farm Economics. 48 (1966): 711-727.

McCalla, A.F. "Wheat and the Price Mechanism: Or Duopoly Revisited and Abandoned." Seminar on Wheat: Papers and Summary of Discussions of the 1969-70 Seminar Series. Dept. of Agricultural Economics, University of Manitoba. Occasional Series No. 2. October 1970.

McKinzie, L.D., III. "An Econometric Model of the Canadian Wheat Market." M.S. Thesis, Purdue University. 1979.

Pindyck, R.S. "Optimal Economic Stabilization Policies under Decentralized Control and Conflicting Objectives." Institute of Electrical and Electronic Engineers. Vol. AC-22, No. 4. August 1977.

Rausser, G.C., E. Lichtenberg, and R. Lattimore. "Developments in Theory and Empirical Applications of Endogenous Governmental Behavior." G.C. Rausser (ed.) New Directions in Econometric Modelling and Forecasting in U.S. Agriculture. Amsterdam: North Holland Press, 1983.

Rausser, G.C. and J.W. Freebairn. "Estimation of Policy Preference Functions: An Application to U.S. Beef Import Quotas." Review of Economics and Statistics. 56 (1974): 437-449.

Rojko, A.S., H. Fuchs, P. O'Brien, and D. Regier. Alternative Futures for World Food in 1985. ECSC/ USDA. Foreign Agricultural Economics Report 146, 149, 151. 1978.

Schmitz, A. and L. Bawden. The World Wheat Economy: An Empirical Analysis. Giannini Foundations Monograph No. 32. University of California, Berkeley. 1973.

Takayama, T. and G.C. Judge. Spatial and Temporal Price and Allocation Models. Amsterdam: North Holland Press, 1971.

Taplin, J.H. "Demand in the World Wheat Market and the Export Policies of the United States, Canada, and Australia." Ph.D. Thesis. Cornell University. 1969.

U.S. Department of Agriculture. "Analysis of American Agricultural Movement Proposal." Issue Briefing Paper. Office of Governmental and Public Affairs. March 3, 1978.

U.S. Department of Agriculture. Foreign Agricultural Service. Grain Data Base.

U.S. Department of Agriculture. Wheat Situation. ERS/ USDA. Various issues: 1960-1980.

Part III

Market Structure and Price Instability in International Agricultural Trade

9
Domestic Price Policies and International Rice Trade

Alexander H. Sarris and John W. Freebairn

INTRODUCTION

World trade in rice is influence significantly by national rice policies. Falcon and Monke (1980) estimate that various nontariff barriers affect 93 percent of world rice imports and 76 percent of exports. Policies in many of the developing countries result in lower prices to both producers and consumers lower than those in the world market. Most of these policies are designed to keep the cost of staple food low. By contrast, in the developed countries, most price policies involve maintaining domestic prices above world levels in order to support the income levels of domestic producers. There are, however, a few countries (including the United States) that currently do not actively seek to set domestic prices different from world levels.

The most important feature of the national rice policies of many of the exporting and importing countries is that they tend partially to insulate domestic prices from volatile international prices. For example, during the 1970-1980 period, while the coefficient of variation of world rice prices (as represented by the export price of Thai rice, 5 percent broken, FOB Bangkok) exceeded 40 percent, the coefficient of variation of domestic prices in many countries was less than 15 percent.

This paper reports an assessment of the probable effects of potential changes in national rice policies on both the average levels and the variability of international rice prices. We analyze movements

Alexander H. Sarris, University of Athens, Greece and John W. Freebairn, La Trobe University, Australia.

toward less restricted as well as more restricted trade.

We use throughout a model developed by Sarris and Freebairn. In that model, domestic prices are regarded as the result of domestic welfare considerations and are derived from the solution of a welfare optimization problem. The welfare function includes producer surplus, consumer surplus, government payments, and variability of domestic prices faced by producers and consumers. These are all weighted by implicit welfare weights which are subsequently inferred from observed prices. Solution of the optimization problem for each country produces an excess demand function with world price as one of the causal variables. Combining the excess demand function for all countries in a Cournot equilibrium fashion results in a market-clearing identity that yields world prices and trade patterns. Analytical solutions are obtained for average values and standard deviations of domestic prices in all countries. Different policies are assessed by changing the welfare weights or the parameters of the rice supply and demand functions and recomputing expected values and standard deviations of prices.

The rest of the paper is organized as follows. In the second section we outline the model used. In the third section we exhibit information on prices, quantities, and supply and demand price elasticities for rice in the principal importing and exporting countries that is required to apply the model. Estimates of the effects of selected policy changes are reported in the fourth section. A final section contains some concluding comments.

DOMESTIC PRICE POLICY FORMATION IN ONE COUNTRY

In this section we briefly outline the model. First of all consider one of m countries trading in the relevant commodity. The domestic supply and demand curves of this country for the particular commodity are assumed to be linear with additive disturbances.

$$S_t = -a + b \, p_{st} - u_t \tag{1}$$

$$D_t = c - d \, p_{dt} + v_t, \tag{2}$$

where p_{st} and p_{dt} are prices faced by domestic producers and consumers, respectively, in year t; u_t and v_t are zero mean independent random variables with variances denoted by σ_u^2 and σ_v^2, respectively; and b, c, and d are greater than 0.

Assume that, in year t, the world price for the

commodity is given by p_{wt} and that every trading country takes it as exogenous. The welfare gain of domestic producers in year t, GP_t, from obtaining price p_{st} rather than p_{wt} is:

$$GP_t = \frac{1}{2}(p_{st} - p_{wt}) \left[S_t(p_{st}) + S_t(p_{wt}) \right] \tag{3}$$

Similarly, the welfare gain by consumers in year t, GC_t, from obtaining price p_{dt} rather than p_{wt} is:

$$GC_t = \frac{1}{2}(p_{wt} - p_{dt}) \left[D_t(p_{dt}) + D_t(p_{wt}) \right]. \tag{4}$$

The monetary gain to the government treasury in year t, GG_t, from maintaining these deviations from world price is:

$$GG_t = p_{dt} \, D_t(p_{dt}) - p_{st} \cdot S_t(p_{st}) -$$

$$p_{wt} \cdot \left[D_t(p_{dt}) - S_t(p_{st}) \right]. \tag{5}$$

Suppose that prices p_{st} and p_{dt} (or the policies that result in these prices) are set every year so as to maximize the following welfare objective function:

$$W_t = w_p \, GP_t + w_c \, GC_t + w_g \, GG_t - \frac{z_p}{2}(p_{st} - p_{d_s}^e)^2$$

$$- \frac{z_c}{2}(p_{dt} - p_d^e)^2. \tag{6}$$

In equation (6), w_i (i = p, c, g) are implicit welfare weights for producers, consumers, and government expenditures in the commodity. The last two terms incorporate domestic preferences for producer and consumer price stability respectively. The terms p_s^e and p_d^e are expected values of the domestic producer and consumer prices, respectively.

After substitution of expressions (1)-(5), the expression in equation (6) is quadratic in p_{st} and p_{dt}. Maximizing W_t with respect to these two prices, we obtain the following expressions for optimal domestic price policies for producers p_{st} and consumers p_{dt}:

$$p_{st}^* = \frac{(w_g - w_p)(a + u_t) + w_g b \, p_{wt} + z_p \, p_s^e}{(2w_g - w_p) b + z_p} \tag{7}$$

$$p_{dt}^* = \frac{(w_g - w_c)(c + v_t) + w_g d P_{wt} + z_c p_d^e}{(2w_g - w_c) d + z_c} \qquad (8)$$

The second-order conditions for maximization are satisfied when the denominators in expressions (7) and (8) are positive--an assumption maintained throughout that is not contradicted in the empirical results reported later.

Taking the expected values of expressions (7) and (8), we can solve for p_s^e and p_d^e as follows:

$$p_s^e = \frac{(w_g - w_p) a + w_g b p_w^e}{(2w_g - w_p) b} \qquad (9)$$

$$p_d^e = \frac{(w_g - w_c) c + w_g d p_w^e}{(2w_g - w_c) d} . \qquad (10)$$

By considering the differences $p_s^e - p_w^e$ and $p_d^e - p_w^e$, it can easily be shown that conditions $w_p > w_g$ implies that $p_s^e - p_w^e > 0$ and condition $w_c > w_g$ implies that $p_d^e - p_w^e < 0$. In other words, if producer welfare (as measured by producer surplus), is given a higher social weight than government welfare (as measured by budget outlays), then the average producer price is set above the average world price. Similarly, if consumers' welfare is valued higher than government expenditures, the average consumer price is lower than the average world price.

From (7) and (8) it can be observed that, if $w_p = w_c = w_g$ and $z_p = z_c = 0$, then $p_{st}^* = P_{wt}$ and $p_{dt}^* = P_{wt}$. In other words, if there is no special preference for producers, consumers, or treasury gains and if there is no concern about domestic price stability, the optimal policy for the country is free trade.

Since parameter z_p and z_c do not appear in expressions (9) and (10), price policies aimed strictly toward domestic stability (i.e., when z_p, $z_c \neq 0$) do not affect the average level of domestic prices although they do affect the yearly prices as is indicated in the basic expressions [(7) and (8)].

By substituting expressions (7) and (8) into (1) and (2), respectively, one can derive the excess demand in year t, ED_t:

$$ED_t = D_t(p_{dt}^*) - S_t(p_{st}^*)$$

$$= c(1 - \gamma d) + a(1 - \epsilon b) + v_t(1 - \gamma d) \qquad (11)$$

$$+ u_t(1 - \epsilon b) - (\delta d + \zeta b) P_{wt} - \rho d p_d^e - \xi b p_s^e$$

where

$$\gamma = \frac{w_g - w_c}{(2w_g - w_c)\, d + z_c} \tag{12}$$

$$\varepsilon = \frac{w_g - w_p}{(2w_g - w_p)\, b + z_p} \tag{13}$$

$$\delta = \frac{w_g\, d}{(2w_g - w_c)\, d + z_c} \tag{14}$$

$$\zeta = \frac{w_g\, b}{(2w_g - w_p)\, b + z_p} \tag{15}$$

$$\rho = \frac{z_c}{(2w_g - w_c)\, d + z_c} \tag{16}$$

$$\xi = \frac{z_p}{(2w_g - w_p)\, b + z_p}. \tag{17}$$

Denoting every trading country by a subscript i and setting the sum of all excess demands to zero (assuming m trading countries), we can derive an expression of the world price, p_{wt}:

$$
p_{wt} = \frac{1}{\sum\limits_{i=1}^{m} (\delta_i\, d_i + \zeta_i\, b_i)} \left\{ \sum\limits_{i=1}^{m} \left[c_i (1 - \gamma_i\, d_i) \right. \right.
$$

$$
+ a_i (1 - \varepsilon_i\, b_i) + v_{i}(1 - \gamma_i\, d_i)
$$
$$\tag{18}$$

$$
+ u_{it} (1 - \varepsilon_i\, b_i) - \rho_i\, d_i\, p_{di}^{e}
$$

$$
\left. \left. - \xi_i\, b_i\, p_{si}^{e} \right] \right\}.
$$

Taking the expected value of equation (20) and considering expressions (9) and (10), we can derive the following expression for the expected world price:

$$
p_w^{e} = \frac{\sum\limits_{i=1}^{m} \left(\dfrac{w_{gi}\, c_i}{2w_{gi} - w_{ci}} + \dfrac{w_{gi}\, a_i}{2w_{gi} - w_{pi}} \right)}{\sum\limits_{i=1}^{m} \left(\dfrac{w_{gi}\, d_i}{2w_{gi} - w_{ci}} + \dfrac{w_{gi}\, b_i}{2w_{gi} - w_{pi}} \right)}. \tag{19}
$$

The variance of world price can be found by

taking the expected value of the square of the difference between expressions (18) and (19). Under the rather innocuous and theoretically very convenient assumption that all domestic fluctuations are independent across countries, the variance is given by the following expression:

$$\sigma_w^2 \equiv E(p_{wt} - p_w^e)^2 = \frac{\sum\limits_{i=1}^{m} \left[\sigma_{vi}^2 (1 - \gamma_i \, d_i)^2 + \sigma_{ui}^2 (1 - \varepsilon_i \, b_i)^2\right]}{\left[\sum\limits_{i=1}^{\overline{m}} (\delta_i \, d_i + \zeta_i \, b_i)\right]^2} . \tag{20}$$

Expressions (19) and (20) show explicitly the dependence of the mean and variance of international prices on the various weights that are used implicitly by each country in deriving its domestic price policies.

It is relatively easy to show the following relationships:

$$\frac{\partial p_w^e}{\partial w_{pi}} < 0 \qquad \frac{\partial p_w^e}{\partial w_{ci}} > 0 \qquad (i = 1, \ldots, m) \tag{21}$$

$$\frac{\partial \sigma_w^2}{\partial z_{pi}} > 0 \qquad \frac{\partial \sigma_w^2}{\partial z_{ci}} > 0 \qquad (i = 1, \ldots, m). \tag{22}$$

The partial derivatives of p_w^e with respect to w_{gi} and those of σ_w^2 with respect to w_{pi}, w_{gi}, and w_{ci} ($i = 1, \ldots, m$) are of ambiguous sign.

The relationships in (22) summarize in a nutshell the conventional wisdom about domestic stabilization policies, namely, they tend to destabilize international prices. The relationships in (21) are intuitively clear since a price policy favorable to producers tend to increase domestic production and, hence, to increase exportable surplus (or, equivalently, decrease import demand); a domestic price policy favoring consumers tends to increase world prices because, by decreasing domestic consumer prices, it increases domestic demand and, consequently, the country's excess demand.

Using expressions (7)-(10) and (19) and (20), we can compute the variance of the domestic supply and demand prices as follows:

$$\sigma_{di}^2 \equiv E\left[p_{dit}^* - p_{di}^e\right]^2 = \gamma_i^2 \sigma_{vi}^2 + \delta_i^2 \sigma_w^2 + 2\gamma_i \delta_i$$

$$\frac{(1 - \gamma_i d_i) \sigma_{vi}^2}{\frac{1}{m}\sum_{k=1}^{m}(\delta_k d_k + \zeta_k b_k)} \tag{23}$$

$$\sigma_{si}^2 \equiv E\left[p_{sit}^* - p_{si}^e\right]^2 = \varepsilon_i^2 \sigma_{ui}^2 + \zeta_i^2 \sigma_w^2 + 2\varepsilon_i \zeta_i$$

$$\frac{(1 - \varepsilon_i b_i) \sigma_{ui}^2}{\frac{1}{m}\sum_{k=1}^{m}(\delta_k d_k + \zeta_k b_k)}. \tag{24}$$

From equations (23), (24), and (12)-(17), it can be seen that large values of z_p and z_c render the parameters γ_i, δ_i, ε_i, and ζ_i small, thus stabilizing domestic prices.

Notice in the model described above that the domestic prices used are not necessarily the ones observed in the market. They are the prices that would have resulted if all relevant policies could be translated in corresponding price terms. For example, a quota on imports or a system of restitution payments could be translated to an equivalent price increase for producers. The model is of the Cournot equilibrium type implying that, in designing their domestic policies, trading countries do not consciously consider the price policies (i.e., the w and z weights) of other countries.

The model, being a partial equilibrium one, assumes no interactions for other commodity markets and no constraints on government expenditures. Although these are realistic considerations, they do not subtract much from the later conclusions.

SOURCES OF DATA

For the policy analysis, the world rice market is disaggregated into 29 countries or regions. The disaggregation isolates the principal exporting and importing countries, and it is one for which a reasonable set of data could be obtained.

Table 1 exhibits the basic data. They are designed to characterize the world rice situation circa 1980. Average quantities produced and traded for 1978 through 1980 are taken from the United Nations Food and Agriculture Organization production and trade yearbooks. Average utilization is obtained by adding average production and average net imports. The figures illustrate the dominance of Asian countries, the thinness of world rice trade (only 5 percent of world production is exported), and the lack of dominance of the world market by any one country.

Consumer and producer prices reflect the effects of current policies. The average prices refer to perceived longer run distributional preferences while the standard deviations refer to perceived preferences for price stability. The world average price of $378 (U.S.) per metric ton is the average price of Thai white rice exports (5 percent broken, f.o.b. Bangkok) for 1978 through 1980 reported in Palacpac. The standard deviation of this price was calculated for the deflated series over the 1970-1980 period. All other prices are specified relative to this one. The comparison of prices is complicated by product heterogeneity, transport costs, and marketing charges --all of which are ignored here. Fortunately, Petzel and Monke (1980) have shown that there exists a high degree of correlation among prices of rice with different qualities and from different countries in the international market. Prices for the different countries are based on a number of sources for direct price data and indirect data on national policies (Palacpac (1982); Falcon and Monke (1980); Tyers and Chisholm (1982); Bale and Lutz (1981); and miscellaneous individual country yearbooks and journal articles not shown in the reference list). Considerable judgment was involved, and both the mean and standard deviation estimates can be defended only as first approximations.

Many of the countries, including exporters and importers, maintain domestic prices substantially different from international prices (Japan, Malaysia, Taiwan, and South Korea) or effectively stabilized (Burma, India, Indonesia, and Thailand).

Sources for the elasticities are recent journal articles (not shown in the reference list) and the world grain-oilseed-livestock (GOL) projection model of Rojko et al (1978). Short-run supply elasticities were computed from long-run elasticities by dividing by three; short-run demand elasticities were obtained from long run figures divided by two.

Estimates of the standard deviations of the additive error terms of the supply equations are based on our estimates of the coefficients of variation of yields during the period 1970-1980 times the base period production figures. These estimates are adjusted slightly in the base run to achieve internal consistency of the model.

RESULTS

Given the base data exhibited in Table 1, the

implicit welfare weights can be computed by solving
numerically the simultaneous nonlinear equations (9),
(10), (23), and (24) for m countries for the average
prices and standard deviations exhibited in section 2.
Table 2 exhibits the results of this exercise (the
normalization $w_g + w_p + w_c = 3$ was assumed so that,
under free trade, $w_g - w_p = w_c = 1$ and $z_p = z_c = 0$).
Interestingly, the values for the z_p and z_c weights are
very small, reflecting the great sensitivity of the
model to small changes in the z parameters. It can be
observed that, for several countries, $w_p < 1$ and $w_c > 1$.
This, in essence, means that those countries are taxing
producers but favor consumers in their rice policies.
      Given the structure of the model, the impact on
international prices of changing the policies of any
one country or group of countries can be simulated.
In Table 3, we report the results of a series of
experiments in which individual countries or groups of
countries are assumed to liberalize trade. This is
simulated in the model by setting $w_g = w_p = w_c = 1$ and
$z_p = z_c = 0$ for the liberalizing countries and leaving
all of the other country weights at their base values.
      The results are interesting. If all countries
liberalized trade, the average world rice price would
decline by 16.3 percent and its standard deviation
would decline by 61 percent. This result is in stark
contrast with the result for wheat reported for the
same model by Sarris and Freebairn (1983). There,
world trade liberalization lead to an increase of
average world wheat price.
      The reason for this result is that, in rice, the
major producing and trading countries are developing
countries with large populations. They tend to pursue
food policies that keep the price of rice to both
producers and consumers quite low. In wheat, by
contrast, the major trading countries are developed
countries that tend to keep their grain prices high in
order to support domestic producers.
      Among the major rice trading nations, Indonesia,
Thailand, India, and Burma seem to have the greatest
impact via their policies on world prices; this is a
result of their large production and consumption
coupled with their isolationist domestic rice policies.
      Interestingly enough, the United States, despite
its dominant trade position, would not have much
influence on world prices if it instituted restrictive
domestic rice policies. This was simulated by setting
for the United States w_p equal to 1.05 and z_p equal to
0.02. This results in an almost perfectly stable
average domestic rice price in the United States of
$908 per metric ton but an international average
price of $375.1 (U.S.) per metric ton (standard
deviation, 164.7) which represents changes from the
base values close to 1 percent. Even smaller changes

TABLE 1
Basic Rice Data Used in the Model[a]

			Country or Region		
	World	Bangladesh	Burma	China / North Korea	Hong Kong / Singapore
Average					
Production[e]	251.81	12.96	6.88	93.82	0.
Utilization[e]	g	13.23	6.33	92.32	.53
Net Imports[f]		.27	-.55	-1.50	.53
Producer Price[f]	378[h]	340	200	378[i]	
Standard Deviation of Producer Price[f]	162	95	25		
Average Consumer Price[f]		340	200	162[i]	378[i]
Standard Deviation of Consumer Price[f]		95	25	162[i]	162[i]
Price Elasticity					
Supply		.06	.13	.17[j]	.15[j]
Demand		.20	.20		
Standard Deviation of Supply[e]		1.51	1.51	.90[k]	0[k]

TABLE 1 Continued
Basic Rice Data Used in the Model[a]

| | Country or Region | | | | |
	India	Indonesia	Japan	Malaysia	Middle East[b]
Average					
Production[e]	48.60	17.73	9.29	1.23	.98
Utilization[e]	48.29	19.66	8.86	1.51	2.51
Net Imports[f]	- .31	1.93	- .43	.28	1.53
Producer Price[f]	300	300	2,150	450	378
Standard Deviation of Producer Price[f]	30	60	150	100	100
Average Consumer Price[f]	300	300	1,500	378	378
Standard Deviation of Consumer Price[f]	50	60	150	100	100
Price Elasticity					
Supply	.06	.10	.05	.03	.10
Demand	.20	.30	.06	.20	.10
Standard Deviation of Supply[e]	6.99	2.51	.92	.09	.13

TABLE 1 Continued
Basic Rice Data Used in the Model[a]

	Country or Region				
	Pakistan	Phillippines	South Korea	Sri Lanka	Taiwan
Average					
Production[e]	3.11	4.93	4.53	1.26	2.07
Utilization[e]	2.10	4.79	4.90	1.46	1.76
Net Imports[f]	-1.01	-.14	.37	.20	-.31
Producer Price[f]	300	300	1,170	250	550
Standard Deviation of Producer Price[f]	60	50	150	100	150
Average Consumer Price[f]	300	300	1,170	250	550
Standard Deviation of Consumer Price[f]	70	50	150	100	150
Price Elasticity					
Supply	.06	.03	.06	.07	.07
Demand	.20	.20	.15	.20	.20
Standard Deviation of Supply[e]	.15	1.09	.82	.25	.16

TABLE 1 Continued
Basic Rice Data Used in the Model[a]

		Country of Region			
	Thailand	Vietnam Laos	Other Asia[l]	Egypt	Nigeria
Average					
Production[e]	11.21	6.53	4.45	1.57	.61
Utilization[e]	8.89	6.90	4.83	1.42	1.04
Net Imports[e]	- 2.32	.37	.38	- .15	.43
Producer Price[f]	280		378	200	378
Standard Deviation of Producer Price[f]	50		100	50	100
Average Consumer Price[f]	300	378[i]	378	150	378
Standard Deviation of Consumer Price[f]	55	162[i]	100	50	100
Price Elasticity					
Supply	.14		.07	.08	.12
Demand	.22	.20[j]	.20	.13	.20
Standard Deviation of Supply[e]	.99	.47[k]	.50	.12	.05

TABLE 1 continued.
Basic Rice Data Used in the Model[a]

	Country or Region				
	Other Africa	Brazil Peru	Other South America	United States	Cuba
Average					
Production[e]	3.16	5.96	2.37	4.04	.30
Utilization[e]	4.85	6.44	1.81	1.54	.49
Net Imports[e]	1.69	.48	-.56	-2.50	.19
Producer Price[f]	378	378	378	378	
Standard Deviation of Producer Price[f]	100	120	120	162	
Average Consumer Price[f]	378	378	378	378	378[i]
Standard Deviation of Consumer Price[f]	100	120	120	162	162[i]
Price Elasticity					
Supply	.12	.05	.06	.08	
Demand	.20	.10	.20	.10	.10[j]
Standard Deviation of Supply[e]	.37	.43	1.09	.19	.07[k]

TABLE 1 Continued
Basic Rice Data Used in the Model[a]

		Country or Region			
	Other North and Central America	European Community[c]	Other Europe[l]	Soviet Union	Australia
Average					
Production[e]	.99	.67	.54	1.58	.45
Utilization[e]	1.18	.88	1.11	2.09	.26
Net Imports[f]	.19	.21	.57	.51	-.19
Producer Price[f]	378	460	378		378
Standard Deviation of Producer Price[f]	100	50	100		162
Average Consumer Price[f]	378	460	378	378[i]	378
Standard Deviation of Consumer Price[f]	100	50	100	162[i]	162
Price Elasticity					
Supply	.05	.07	.07	.15[j]	.03
Demand	.15	.15	.15		.08
Standard Deviation of Supply[e]	.13	.13	.07	.29[k]	.07

Continued.....

TABLE 1 Concluded.
Basic Rice Data Used in the Model[a]

<u>Source</u>: Computed

[a]Basic data designed to characterize the world rice situation circa 1980.

[b]Iran, Kuwait, Saudi Arabia, United Arab Emirates, and Yemen.

[c]Belgium, Denmark, England, Federal Republic of Germany, France, Ireland, Italy, Luxemburg, The Netherlands.

[d]Derived on a milled-rice basis; paddy rice has been multiplied by .65 to obtain figures in milled equivalents.

[e]Million metric tons.

[f]Dollars (US) per metric ton.

[g]Blanks indicate data are not needed.

[h]Average international price; this price and its standard deviation are used for normalization.

[i]For these countries or regions, only excess demands are considered; the world price and standard deviation are used for normalization.

[j]Elasticities of excess demand (or supply).

[k]Standard deviation of net imports (or exports) only.

[l]Derived as residual.

TABLE 2
Implicit Welfare Weights of Rice-Trading Countries Consistent with Current Domestic
Price Policies

	Welfare Weights			Producer Variability	Consumer Variability
	Government Expenditure $(w_g)^c$	Producers Surplus (w_p)	Consumers Surplus (w_c)	(z_p)	(z_c)
Bangladesh	.9948	.9881	1.017	.00161	.00563
Burma	.9806	.8643	1.1551	.02615	.03512
China, North Korea	1.0	1.0	1.0	0	0
Hong Kong, Singapore	1.0	1.0	1.0	0	0
India	.9880	.9726	1.0394	0.4522	.07290
Indonesia	.9830	.9574	1.0596	.01008	.03436
Japan	1.0003	1.0443	.9554	.00013	.00001
Malaysia	.9982	1.0035	.9982	.00001	.00049
Middle East[d]	1.0	1.0	1.0	.00016	.00041
Pakistan	.9880	.9726	1.0394	.00104	.00189
Philippines	.9858	.9772	1.0370	.00124	.00722
South Korea	1.0215	1.0607	.9178	.00001	-.00001
Sri Lanka	.9778	.9444	1.0779	.0002	.00082
Taiwan	1.0141	1.0352	.9507	.00003	.00001
Thailand	.9973	.9496	1.0531	.01211	.01272
Vietnam, Laos	1.0	1.0	1.0	0	0
Other Asia	1.0	1.0	1.0	.00049	.00158
Egypt	.9628	.8914	1.1457	.00137	.00277
Nigeria	1.0	1.0	1.0	.00012	.00034
Other Africa	1.0	1.0	1.0	.00061	.00159

Continued ....

TABLE 2 Concluded.
Implicit Welfare Weights of Rice-Trading Countries Consistent with Current Domestic
Price Policies

| | Welfare Weights | | | | |
	Government Expenditure $(w_g)^c$	Producers Surplus (w_p)	Consumers Surplus (w_c)	Producers Variability (z_p)	Consumer Variability (z_c)
Brazil, Peru	1.0	1.0	1.0	.00028	.00060
Other South America	1.0	1.0	1.0	.00015	.00034
United States	1.0	1.0	1.0	0	0
Cuba	1.0	1.0	1.0	0	0
Other North and Central America	1.0	1.0	1.0	.00008	.00029
European Community[b]	1.0050	1.0169	.9781	.00022	.00064
Other Europe	1.0	1.0	1.0	0	0
Soviet Union	1.0	1.0	1.0	0	0
Australia	1.0	1.0	1.0	0	0

Source: Computed

[a]Iran, Kuwait, Saudi Arabia, United Arab Emirates, and Yemen.

[b]Belgium, Denmark, England, France, Federal Republic of Germany, Ireland, Italy, Luxemburg, The Netherlands.

[c]Notation corresponds to that in the text.

TABLE 3
Results of Trade Liberalization Simulations[a]

Liberalizing Country or Region	Average World Price dollars (US)	Change from Base percent	Standard Deviation of World Price dollars (US)	Change from Base percent
All countries	316.3	-16.3	63.1	-61.0
Japan	384.0	1.6	161.6	- .2
Indonesia	363.0	- 4.0	122.7	-24.3
Thailand	370.3	- 2.0	139.7	-13.8
Egypt	375.1	- .8	158.0	- 2.5
European Community[b]	378.2	0	161.1	- .6
India	353.3	- 6.5	101.2	-37.5
Burma	363.4	- 3.9	136.8	-15.6
Philippines	375.8	- .5	154.0	- 4.9
South Korea	383.0	1.3	161.6	- .2
Bangladesh	375.1	- .8	149.6	- 7.7

Source: Computed.

[a]Circa 1980.

[b]Belgium, Denmark, England, France, Federal Republic of Germany, Ireland, Italy, Luxemburg, The Netherlands.

are projected if the nine-member European Community
adopts more restrictive rice price-support policies.
    China is one of the largest net exporting
countries. Yet, we know very little about its
internal rice policies. To obtain an idea of China's
impact on world trade, we double its export supply
price elasticity. The result is a 1.8 percent
decrease of the standard deviation of world rice price
(the mean does not change because we do not change the
average quantity exported). It would appear that
China can influence world rice prices by the price
responsiveness of its export policy.
    Much smaller changes in the variability of world
prices are found with the excess demand price
elasticities of the other centrally planned countries
(Soviet Union, Cuba, and Vietnam, Laos).

SUMMARY

    The results of this paper have illustrated the
interaction of various national rice policies in the
international market. It was shown that the result of
these policies in the aggregate is an artificial
increase in the average world price and a substantially
increased instability. The coefficient of variation
of the world rice price, which is currently about 43
percent, would drop to 20 percent under a completely
free rice trade regime.
    It was also shown that the countries which can
influence world prices the most are the major producers
and, to a lesser extent, the large exporters which are
relatively small producers.
    Stability in the world rice market can be
achieved in ways other than free trade. One such way
is the institution of national or regional buffer
stocks. This type of policy has been ignored
completely in this paper because stocks are not part
of the model; but it, also, can have significant
stabilizing influences. The results of buffer
simulation exercises, nevertheless, could be compared
with the results here in order to assess the relative
benefits of different policy choices.

REFERENCES

Bale, M. D., and E. Lutz. "Price Distortions in
    Agriculture and Their Effects: An International
    Comparison." American Journal of Agricultural
    Economics 63(1981): 8-22.
Falcon, W. P., and E. A. Monke. International Trade
    in Rice. Stanford, California: Stanford Food
    Research Institute Studies No. 17, 1980.

Palacpac, A. C. World Rice Statistics. Los Banos, The
    Philippines: International Rice Research
    Institute, 1982.
Petzel, T. E., and E. A. Monke. "The Integration of
    the International Rice Market." Stanford,
    California: Stanford Food Research Institute
    Studies No. 17, 1980.
Rojko, A., D. Regier, P. O'Brien, A. Coffing, and
    L. Bailey. Alternative Futures for World Food
    in 1985. U. S. Department of Agriculture Foreign
    Economic Reports Vols. 1, 2, and 3, Nos. 146, 149,
    and 151, 1978.
Sarris, A. H., and J. W. Freebairn. "Endogenous Price
    Policies and International Wheat Prices."
    American Journal of Agricultural Economics. 65
    (1983): 214-224.
Tyers, R., and A. Chisholm. "Agricultural Policies in
    Industrialized and Developing Countries and
    International Food Security." Paper presented at
    annual meeting of the Australian Agricultural
    Economic Society, Melbourne, 1982.
United Nations, Food and Agriculture Organization.
    FAO Production Yearbook. Rome, 1978-1980.
United Nations. FAO Trade Yearbook. Rome, 1980.

10

The Sugar Market Policy of the European Community and the Stability of World Market Prices for Sugar

Peter M. Schmitz and Ulrich Koester

INTRODUCTION

The European Community (EC) insulates most of the domestic agricultural markets from the world markets by imposing variable levies on imports and by paying export restitutions for exports. Some researchers point out that the system of variable import levies being applied to most of the EC agricultural products increases instability in world market prices (Johnson (1975), Shei and Thompson (1977), Tangermann (1978, 1981), Zwart and Meilke (1979), Bale and Lutz (1979, 1981), Sampson and Snape (1980), Josling (1980, 1981), Lutz and Bale (1980) and Svedberg (1981)). Although this view seems widely accepted, it is nevertheless investigated in some detail in this paper.

The problem is first considered theoretically. After a short summary of the accepted state of knowledge, assumptions and relevant theories are discussed and a market model is presented. The considerations are quite general and hold for any variable levy system. As the general conclusions with respect to destabilizing effects are inconclusive, a definite answer for a specific market can only be given on the basis of an empirical investigation. This is done for the sugar market.

The EC sugar market is chosen for several reasons: First, the EC has developed to one of the largest sugar producing and exporting regions. EC sugar production accounted for 15.8 percent of world sugar production in 1980. The export share reached 21.7 percent. EC producer prices were more than double world market

Peter M. Schmitz and Ulrich Koester, Institut für Agrarpolitik und Marktlehre der Christian-Albrechts-Universität, Kiel, Germany.

prices (see Appendix Table A 1). Hence, it may be expected that liberalizing the EC sugar market policy might have a significant effect on the instability of world market prices. Second, the policy instruments of the EC sugar market include a quota system in addition to the variable levies and variable export restitutions that exist for most other EC agricultural products. It is of special interest to examine if such a quota system affects the stability of world sugar prices. Third, sugar is one of the few agricultural products where the EC is competing directly with producers in developing countries, because sugar can be produced from sugar beets as in the EC or from sugar cane as in the tropics.

THE EC IMPORT LEVY/EXPORT RESTITUTION SYSTEMS AND THE INSTABILITY OF WORLD MARKET PRICES

The Present State of Knowledge

The EC insulates its domestic markets from the world market by setting threshold prices for foreign competitors. The difference between the threshold price and the world market price is made up by levies or export restitutions. If world market prices change, it is the levy or the restitution that is allowed to vary but the threshold price is kept constant. As the threshold price determines the domestic producer and consumer prices, domestic consumption will not adjust to a variation of world market prices. Thus, EC market regulations make EC import demand or export supply completely inelastic with respect to world market prices. Consequently, the world market demand and supply elasticities are smaller than they would be if EC markets were liberalized. An exogenous shock to supply, such as bad weather, would cause greater changes in world market prices with an organized EC market than with a liberalized EC market. Although this argument seems to be plausible and even consistent with economic theory, some of the critical assumptions have to be discussed.

The Relevance of the Reference System

Evaluations of policy instruments often differ because a given instrument has been compared to alternatives. The variable levy system could be analyzed with reference to an ad valorum tariff giving the same protection as the variable levy system over the same time period, but differing in abnormal world market conditions. It could also be analyzed with reference to a free trade situation. This alternative seems reasonable, as the EC is blamed for its

protectionist agricultural policy. We may ask whether
by-products of this policy tend to destabilize or
stabilize the world market.

The Origin and Magnitude of Exogenous Shocks

Exogenous shocks may originate from supply or
demand. As we are primarily analyzing agricultural
supply it seems reasonable to concentrate on supply-
shocks as normally done in other research. However,
we do not consider the magnitude of exogenous shocks as
given. The size of the exogenous shock may depend on
the policy system. If, for example, the EC has a
variable levy system, it produces more than under free
trade and the rest of the world produces less. Hence,
fluctuations in world production might be different
due to the EC variable levy system. However, this
depends on the kind of assumptions being made about the
form of the disturbances.

(a) <u>Additive or multiplicative disturbances</u>. If
disturbances are additive, supply curve shifts to the
right or left are parallel. In the case of
multiplicative disturbances the supply curve shifts are
not parallel (Turnovsky (1978b) and Ryll (1981)). The
assumption about the form of the disturbances is
important for the results. The amount produced has no
relevance for the variance of production if
disturbances are additive as the variance of production
is equal to the variance of the disturbances. However,
the variance of production depends on the amount
produced if disturbances are multiplicative. To show
this we assume that total production q is equal to area
harvested A times yields y. So $q = A \cdot y$. If the area
is deterministic and only the yields are stochastic
the variance of q will be, $\sigma_q^2 = A^2 \cdot \sigma_y^2$, where $\sigma_y^2 =$
variance of y. This clearly shows that the assumption
about the disturbances may affect the findings if the
variable A differs for the two alternatives being
compared.

The authors previously cited either explicitly or
implicitly assumed that disturbances were additive.
We advocate, however, the assumption that disturbances
are multiplicative as the area is planned and only the
yields are stochastic.

(b) <u>Dependent or independent disturbances</u>. The
exogenous shock in total world production is due to
supply fluctuations in the EC and in the rest of the
world. In general, we may write for the variance in
total world production σ^2:

$$\sigma^2 = \sigma_1^2 + \sigma_2^2 + 2\ r\ \sigma_1\ \sigma_2, \tag{1}$$

where

σ_1^2 = variance in EC production,

σ_2^2 = variance in rest of world production,

r = coefficient of correlation between random disturbances,

σ_2 = standard deviation of EC production, and

σ_1 = standard deviation of rest of world production.

If we assume that disturbances are independent r = 0, then the variance in total production is:

$$\sigma^2 = \sigma_1^2 + \sigma_2^2 . \tag{2}$$

As dependence in the disturbances (that is, $r \neq 0$ cannot be precluded a priori, we advocate incorporating this case in the analysis.

(c) <u>Measure of instability</u>. It is quite common to accept the variance as a measure of instability. This was done, for example, by Bale and Lutz (1979) who presented the most formal and consistent approach to the problem. However, the variance may be a misleading criteria of instability if the instability of two variables with different means are compared (Ryll (1981)). As planned production in the EC is higher with the present levy system than it would be under free trade, the same variance or standard deviation may have quite different implications. The coefficient of variation would be a better measure of instability.

(d) <u>Functional form of supply and demand functions</u>. Research findings concerning the problem under consideration have, so far, been derived from linear demand and supply functions. As this may restrict the generality of the findings, more general functions are applied in this study.

(e) <u>Relationship between production and supply</u>. It is conceivable that, even with greater production instability, world market prices are more stable as stockpiling may have counter-balancing effects. Sarris (1982) showed convincingly that higher production instability may be completely compensated by stabilizing storage activities if the amount of uncertainty in the market is not affected by policy activities. Hence, in comparing alternative policy systems the effect on the amount of uncertainty in the market has to be investigated. If a given policy strategy leads to a substitution of privately held stocks by public ones the storage policy of the public institution has to be analyzed.

In addition to weather, world market price instability may be caused by production changes due to price expectations. Analysis of instability due to price expectations has been neglected. As the EC has stable domestic prices EC production does not react to cyclical world market prices. As a result the EC levy system may have a stabilizing effect on world market prices. This is taken into consideration.

STATIC MODEL

The Effect of the Variable Levy System on Instability in World Production

Instability in world production is measured by the coefficient of variation. In deriving the model it is assumed that disturbances are multiplicative and that functions in EC production and rest-of-world production might be independent. From equation (1) the formula for the coefficient of variation can be derived by dividing both sides of the equation by q^2, the square of world production. This results in:

$$v(q)^2 = s_1^2 v(q_1)^2 + s_2^2 v(q_2)^2 + s_1 s_2 2rv(q_1)v(q_2) \quad (3)$$

where

$v(q)$ = coefficient of variation of world production,

$v(q_1)$ = coefficient of variation of EC grain production,

$v(q_2)$ = coefficient of variation of rest-of-world production,

s_1 = share of EC in world production,

s_2 = share of rest-of-world production, and

r = coefficient of correlation between random disturbances.

Now assume that reducing the EC rate of protection will affect the shares of EC and rest-of-world production, but will not change the coefficient of variation. This assumption implies multiplicative disturbances. Equation (3) has to be totally differentiated. This yields:

$$dv(q) = \frac{v(q_1)^2}{v(q)} s_1 ds_1 + \frac{v(q_2)^2}{v(q)} s_2 ds_2$$
$$+ \frac{s_2 rv(q_1)v(q_2)}{v(q)} ds_1 + \frac{s_1 rv(q_1)v(q_2)}{v(q)} ds_2 \quad (4)$$

If reallocation of world production from the EC to the rest-of-world by lowering the EC rate of protection were to decrease the variability in world production, it must hold that:

$$d \, v(q) < 0 \quad (5)$$

As $s_1 + s_2 = 1$ \hfill (6)

it always holds that:

$$ds_1 = - ds_2 \quad (7)$$

Taking equations (5) and (7) into account we can derive from (4)

$$\frac{v(q_2)}{v(q_1)} < \frac{s_1 v(q_1) + s_2 rv(q_2)}{s_2 v(q_2) + s_1 rv(q_1)} \quad (8)$$

If it should turn out that fluctuations in EC production and fluctuations in the rest of world production are independent, $r = 0$. In this case, the necessary condition is:

$$\frac{v(q_2)^2}{v(q_1)^2} < \frac{s_1}{s_2} \quad (9)$$

or

$$v(q_2)^2 s_2 < s_1 \, v(q_1)^2 \quad (10)$$

Equations (8) and (9) state that a reduction of the rate of protection for EC producers may destabilize world production even if the coefficient of variation is lower for the EC than for the rest-of-world.

STOCHASTIC MODEL

The Effect of the Variable Levy System on the Instability of World Market Prices

In the following it is assumed that fluctuations in production are not offset by storage. According to a model specification by Kirschke (1983) the following set of equations are formulated:

$$q_{EC}^S = d(p_{EC}, \theta) \qquad \text{EC supply function} \quad (11)$$

$$q_{RW}^S = f(p_w, \Phi) \qquad \text{rest-of-world supply function} \tag{12}$$

$$q_{EC}^D = g(p_{EC}) \qquad \text{EC demand function} \tag{13}$$

$$q_{RW}^D = h(p_w) \qquad \text{rest-of-world demand function} \tag{14}$$

$$d(p_{EC}, \theta) + (p_w, \Phi) - g(p_{EC}) - h(p_w) = 0 \qquad \text{market clearance condition} \tag{15}$$

with

$$q^{S,D} = \text{quantities supplied and demanded,}$$

$$p_{EC,w} = \text{EC price and world market price,}$$

$$\theta = \text{random variable in EC supply,}$$

$$\Phi = \text{random variable in rest-of-world supply, and}$$

$$d,f,g,h = \text{functional notations.}$$

Two trade policies are considered:

$$p_{EC} = p_w = p \qquad \text{free trade policy} \tag{16}$$

$$p_{EC} = \bar{p} > p_w \qquad \text{EC price fixing policy} \tag{17}$$

Following Mood, Graybill and Boes (1974, p. 181) the coefficient of variation of the world market price can be calculated as:

$$v(p) = \frac{[\text{var } p]^{1/2}}{Ep} = \frac{\{\text{var } \theta[p_\theta]^2 + \text{var } \Phi[p_\Phi]^2 + 2\text{cov}(\theta,\Phi)[p_\theta \cdot p_\Phi]\}^{1/2}}{p(\bar{\theta},\bar{\Phi},\bar{p}) + \frac{1}{2}\text{var } \theta[p_{\theta\theta}] + \frac{1}{2}\text{var } \Phi[p_{\Phi\Phi}] + \text{cov}(\theta,\Phi)[p\theta\Phi]} \tag{18}$$

with

$$v(p) = \text{coefficient of variation of world market prices,}$$

$$\text{var} = \text{variance,}$$

$$E = \text{mean,}$$

$$\text{cov} = \text{covariance,}$$

$$p_{\theta,\Phi} = \text{first-order partial derivative of the world price function with respect to } \theta,\Phi,$$

$p_{\theta\theta,\Phi\Phi}$ = second-order partial derivative of the world price function with respect to θ, Φ,

$p_{\theta\Phi}$ = second-order cross partial derivative of the world price function, and

$\overline{p}$ = fixed EC price.

Using the implicit function rule one can arrive at the first-order and second-order derivatives of the world market price function without solving the equation system (11) to (17) for the world market price equilibrium. The corresponding derivatives are calculated in the appendix. The market model allows the inclusion of both different types of disturbances (additive, multiplicative) and different functional forms of the supply and demand curves (linear, nonlinear).

Taking linear functions with additive random variables, for instance (18) reduces to:

$$v(p) = \frac{\{var\ \theta + var\ \Phi + 2\ cov\ (\theta,\Phi)\}^{1/2}}{[d_p + f_p - g_p - h_p] \cdot p(\overline{\theta},\overline{\Phi})}$$

free trade policy   (19)

$$v(\overline{p}) = \frac{\{var\ \theta + var\ \Phi + 2\ cov(\theta,\Phi)\}^{1/2}}{[f_p - h_p] \cdot p(\theta,\Phi,\overline{p})}$$

EC price fixing   (20)
policy

The formulas (19) and (20) confirm in the special case the hypothesis pointed out by the researchers quoted above that the instability with a price fixing policy always exceeds the free trade instability. Moreover, taking the coefficient of variation as an instability measure, the instability increases with increasing protection level. The variance of the world market price is independent of the supply levels and the expected world market price equals the deterministic equilibrium value.

Taking multiplicative random variables and/or nonlinear functions, however, the results differ. Especially, the price standard deviation as the numerator of the equation (18) depends on the supply levels which are determined by the protection level.

$$[var\ p]^{1/2} = \frac{1}{(f_p - h_p)} \cdot [q_{EC}^2 \cdot var\theta\ q_{RW}^2 \cdot var\ \Phi$$
$$+ 2\ q_{EC} \cdot q_{RW} \cdot cov(\theta,\Phi)]^{1/2}$$

EC price fixing policy   (21)

As the EC price support policy leads to a higher share
of EC production and a lower share of rest-of-world
production the standard deviation may increase or
decrease depending on the production shares, the size
of variance, and covariance terms. It may be that the
decrease in the standard deviation is greater than the
decrease in the expected world market price due to an
increase in protection. Hence, the price fixing policy
may stabilize the world market price compared with the
free trade situation. Only empirical testing can
therefore determine whether EC market regulations
stabilize or destabilize world market prices.

Before presenting the empirical findings for the
world and EC sugar markets, two other aspects of the
model should be pointed out. First, it is assumed that
total production always equals total world supply. The
model did not take into consideration that national
storage policy can stabilize market supply. Hence, the
role of EC storage policy has to be empirically tested.
Second, the model allows us to investigate how weather-
caused exogenous shocks affect supply. However, it
does not take into consideration other causes of
instability and that causes of instability might
influence policy reaction of the EC to compensate for
production instability. In considering the world sugar
market we have to take into account that the lagged
supply response causes price-induced cycles. It may be
hypothesized that the EC neutralizes this source of
instability by not allowing EC sugar beet producers to
respond to world market price fluctuations.

Summing up these theoretical considerations, we
have to state that the hypotheses provided by economic
theory about the effects of EC market organization on
the stability of world market prices are inadequate.
The actual effects have to be investigated empirically.

EMPIRICAL EVIDENCE FOR THE EC SUGAR MARKET

Yield Variability in the EC and in the Rest of the World

Empirical information on the instability of the
world sugar economy are outlined in Table 1. The
Table shows the coefficients of variation for acreage,
yield, production, and prices following the approach
of Cuddy and Della Valle (1978). Weather-caused supply
fluctuations can be the consequence of fluctuations in
acreage as well as in yield. Production economics
determine whether weather conditions may affect the
area cultivated under a given crop. It can be taken
for granted that cane and sugar beet acreage in a given
crop year are completely independent of weather
conditions. This highlights two important points. The
indicator for weather-caused supply fluctuations in

TABLE 1
Instability Indexes of Sugar Production and Sugar Prices in the European Community and the Rest of the World, 1961-1980 [1].

Variable	European Community (9)			Rest of the World		
	Index	Instability Measure	$\bar{R}^2$	Index	Instability Measure	$\bar{R}^2$
1961-1974						
Acreage	3.83	v_L	0.84	3.02	v_L	0.81
Yield	8.50	v	n.s.	3.44	v_L	0.81
Production	7.24	v_L	0.77	4.27	v_L	0.90
Price [2]	-	-	-	61.32	v	n.s.
1967-1980						
Acreage	4.90	v_{LL}	0.80	1.86	v_L	0.94
Yield	10.22	v	n.s.	2.77	v	n.s.
Production	6.26	v_{LL}	0.85	3.81	v_{LL}	0.83
Price	6.36	v_{LL}	0.89	56.57	v_{LL}	0.58

[1] v is the coefficient of variation; v_L is the corrected coefficient of variation (linear trend); v_{LL} is the corrected coefficient of variation (log linear trend).

TABLE 1 (cont'd)

Where n.s. appears, the statistic was not significant at the 1 percent level. R^2 is the coefficient of determination of the trend function adjusted by the degrees of freedom.

2/ Raw sugar price of the International Sugar Agreement for the rest of the world; EC intervention price for the European Community. For the former the time series 1960–1973 has been taken because of the price explosion in 1974.

Source: Calculations based on data from the FAO (ed.), Production Yearbook, various issues (Rome, various years), and Bartens and Mosolff (ed.), Sugar Economy, Berlin 1981/82.

sugar production is yield per hectare; a model to investigate instability has to assume a multiplicative relationship between the disturbances. Table 1 clearly shows that yield per hectare of raw sugar in the EC was less stable than in the rest of the world between 1961 and 1980. Whereas the EC produces sugar only out of sugar beets, the rest of the world mainly produces cane sugar. Therefore Table 1 shows that the yields per hectare from sugar cane are more stable than from sugar beets. The total period has been divided into two overlapping subperiods of 14 years each. This was done to get better information about the second subperiod. The calculated coefficients of variation are much higher for the EC's yield per hectare (10.22 percent) than for the rest of the world's yield per hectare (2.77 percent). The differences in the instability indexes between the EC and the rest of the world are even greater for the later subperiod, during which the instability in world market prices was nearly as great as during the subperiod 1961-1974.

These empirical results, however, do not indicate that world production would be more stable with liberalized EC sugar market policies. If the correlation between EC fluctuations and rest of the world fluctuations is negative, EC sugar policy could help stabilize world market prices while increasing instability of EC sugar production. However, a correlation between the first differences of the two time series - EC yield per hectare and rest-of-world yield per hectare - gave a coefficient of correlation of 0.48 for the total period 1961 to 1980. This value is significant at the 5 percent level. Substituting these parameters in equation (8) above, shows that EC sugar policy leads to an increase in instability of world sugar production.

The price instability effects due to stochastic yield disturbances for different trade policies and different elasticities are shown in Table 2. According to the supply and demand reactions (Grosskopf (1979) and Gemmill (1976)) the instability index decreases between 6.6 and 25.5 percent due to an EC sugar market liberalization. This means that the actual EC sugar policy (1982) destabilizes the world market price.

EC Sugar Stockpiling Policy and the Instability of World Market Prices for Sugar

In general, it holds that the variable levy system as applied in the EC leads to a nearly complete substitution of privately held carryover stocks by public stocks. EC prices are totally disconnected from the world market. Hence, expectations about future world market prices are irrelevant for private

TABLE 2
Price Instability Due to Stochastic Yield Disturbances (percent) [1].

Constellations of price elasticities [2]	Coefficient of variation with EC Price Fixing Policy	Coefficient of Variation with Free Trade Policy	Change of Instability Index
Inelastic reactions within the EC and elastic reactions outside the EC	1.97	1.84	– 5.60
Inelastic reactions within and outside the EC	4.36	3.95	– 9.40
Elastic reactions within and outside the EC	1.97	1.56	–20.81
Elastic reactions within the EC and inelastic reactions outside the EC	4.36	3.25	–25.46

[1] The change of expected world market prices due to EC sugar market liberalization has been taken from Koester and Schmitz, The EC Sugar Market Policy and Developing Countries. "European Review of Agricultural Economics", Vol. 9, No. 2 (1982), p. 190. The coefficients of variation are calculated along the formula (18) assuming $\delta_\theta = 0.1022$, $\delta_\Phi = 0.0277$, $r = 0.4776$ and $p_{\theta\theta} = p_{\Phi\Phi} = p_{\theta\Phi} = 0$ (δ_θ EC yield standard

TABLE 2 (cont'd)

1/ deviation δ_ϕ Rest of the World standard deviation, r coefficiet of correlation).
Random variables are assumed to be multiplicative and supply/demand functions
nonlinear/isoelastic.

2/ Elasticities are given in Table A1 in the Appendix.

Source: Authors calculations along the formula (18).

stockholders. EC exporters receive a subsidy to bridge
the gap between higher EC and lower world market prices
at the time of exports. Hence, it does not pay to
build up privately held EC stocks in periods of low
world market prices and release the stocks in periods
of high world market prices. Consequently, carryover
stocks in the EC are mainly held by each country. This
fact does not necessarily imply a less stabilizing
effect of world stockpiling with the EC levy system
than with free trade. It will only be the case if EC
public stocks are less integrated into the world trade
and stockpiling system and/or if the present EC system
increases the amount of uncertainty in the world
market. If the latter holds true, private stockholders
outside the EC might store less than with free trade.
     The effect of EC stockpiling performance with
respect to stabilizing world markets has to be tested
empirically. Schrader (1982) calculated a regression
analysis between EC production and EC net exports and
between EC net exports and world market prices. He
concluded that EC production instability is highly
correlated with EC export instability. The EC actually
considers the world market to be a residual for the
difference between domestic production and domestic
consumption. The world market price only affects EC
exports with a three-year lag, thus indicating that the
behavior of EC exporters is pro-cyclical.
     It has to be noted that the actual stockpiling
behavior is not necessarily a specific of a variable
levy system. It may well be possible to manage a more
stabilizing stockpiling policy without changing the
levy system. However, with respect to past performance
we have to conclude that EC stocks were not well
integrated in world stocks. It is obvious that the
past stockpiling policy on the EC sugar market
increased the amount of uncertainty on the world
market. Consequently, it is very likely that world
stocks outside the EC are lower due to the past EC
storage policy. Thus, EC stock policy on the sugar
market increased the instability in world market
prices.

The Impact of EC Sugar Market Policy on Price Cycles
in the World Sugar Market

     Our analysis of the effect of EC sugar policy on
world market price instability would not be conclusive
if price cycles on the world market caused by lagged
supply responses were not taken into account. For
some markets this can be considered irrelevant, but
not for the world sugar market (Gemmill (1978)). To
make the analysis more general, however, we first
abstracted from the specifics of the EC sugar market

regulations. This allowed us to derive general conditions for the impact of the EC import levy system on world market price cycles. In the second step we incorporated the specifics of the EC sugar market regulations to reach a specific conclusion for the sugar market.

The dynamic stochastic model for investigating the effect of the EC levy system on price-induced world market supply fluctuations contains the following linear equations (Schmitz and Koester (1981)).

$$q_{EC}^{S}(t) = a_{EC}\overset{*}{p}_{EC}(t) + b_{EC} + \theta(t)$$

EC supply function (21)

$$q_{RW}^{S}(t) = a_{RW}\overset{*}{p}_{w}(t) + b_{RW} + \Phi(t)$$

rest-of-world supply function (22)

$$q_{EC}^{D}(t) = c_{EC} - d_{EC}p_{EC}(t)$$

EC demand function (23)

$$q_{RW}^{D}(t) = c_{RW} - d_{RW}p_{w}(t)$$

rest-of-world demand function (24)

$$\overset{*}{p}(t) = \alpha Ep(t) + (1-\alpha)p(t-1)$$

price expectation function (25)

$$q_{EC}^{S}(t) + q_{RW}^{S}(t) = q_{EC}^{E}(t) + q_{RW}^{R}(t)$$

market clearance condition (26)

Equation (26) is the aggregate price expectation function. The weight α reflects the fact that some producers predict rationally, while others use the naive cobweb forecast procedure (Turnovsky (1978)). With $\alpha = 1$ it becomes the rational price expectation and with $\alpha = 0$ it becomes the static or cobweb price expectation. The asymptotic coefficient of variation can be calculated with the following terms:

$$v(p) = \left[\frac{\text{var } \theta + \text{var } \Phi + 2 \text{ cov}(\theta,\Phi)}{(\Sigma a\alpha+\Sigma)^{2} - (\Sigma a(1-\alpha))^{2}}\right]^{1/2} \cdot \frac{1}{P_{w}}$$

free trade policy (27)

$$v(p) = [\frac{var\ \theta + var\ \phi + 2\ cov(\theta,\phi)}{(a_{RW}\alpha_{RW}+d_{RW})^2 - a_{RW}^2(1-\alpha_{RW})^2}]^{1/2} \cdot \frac{1}{\bar{p}_W}$$

EC price fixing policy   (28)

with

$\bar{p}_W$ = deterministic world market price equilibrium.

Assuming stochastic stability in the sense that the asymptotic coefficient of variation is finite (Turnovsky, 1978a) the market will have the usual cobweb characteristics if $0 \leq \alpha < 1$, while the price will simply fluctuate randomly if $\alpha = 1$. With fixed EC consumer and producer prices and rational price expectations ($\alpha = 1$) a liberalization of the EC sugar policy leads to a decrease of price instability of about 8.5 percent (see Table 3). This percent decrease is similar to the calculations with nonlinear functions and multiplicative disturbances in Table 2. If producers use the naive cobweb forecast the instability level is much higher for both policies. A liberalization, however, increases the coefficient of variation from 11.67 percent to 12.02 percent.

The instability index increases about 3 percent. This marginal destabilizing impact of an EC sugar market liberalization is due to EC sugar producers being allowed to respond to price induced world market cycles. However, these results are purely hypothetical insofar as the EC sugar policy does only insulate the EC consumer prices. In order to receive a specific conclusion from the model we must include the special sugar market regulations.

Table 1 indicates that the instability of the area cultivated with sugar beets was greater for the EC than the area planted with sugar beets and cane outside of the EC. This supports the hypotheses that EC production is linked to changes in world market prices and that because of the special market regime, EC production may respond more to changes in world market prices than it would with free trade. A short analysis of the EC sugar market organization will help to substantiate these hypotheses.

The EC sugar producer faces an individual expected price-sale function as characterized in Figure 1. A high price p(A) is guaranteed for a fixed quantity A (quota A). If a farmer's production is greater than quota A, such as quota B, he gets a reduced price p(B) for the excess quantity. The price reduction depends on the world market price because farmers are supposed to pay up to 60 percent of the export restitutions needed to export the excess quantity. However, the reduction cannot be higher

TABLE 3
Price Instability Caused by Stochastic Yield Disturbances and Producers' Price Expectations (percent)1/

Kind of Price Expectations	Coefficient of Variation with EC Price Fixing Policy	Coefficient of Variation with Free Trade Policy	Change of Instability Index
a) With fixed EC Producer and EC Consumer Prices:			
Static Price Expectation	11.67	12.02	3.00
Rational Price Expectation	4.36	3.99	-8.49
b) With Fixed EC Consumer Prices:			
Static Price Expectation	12.98	12.02	-7.40
Rational Price Expectation	4.23	3.99	-5.67

Source: Authors calculations along the formula (28) and (29).

1/ With inelastic reactions within the EC and outside the EC. The elasticities are taken from Table A1. Random variables are assumed to be additive and supply/demand functions to be linear.

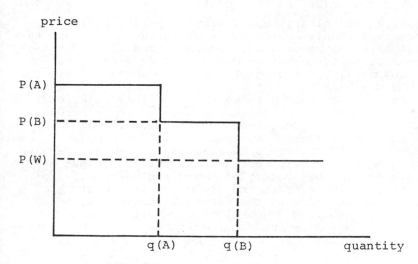

Figure 1   Individual Expected Price Sale Function
            for EC Sugar Beet Producers

than 30 percent of the price p(A). Thus,

$$p(B) \geq 0.7 \, p(A), \tag{30}$$

and

$$p(A) - 0.6[p(A) - p(w)] = p(B), \text{ for } p(A) > p(W) \tag{31}$$

If an individual farmer produces more than q(B), he
receives the world market price p(w) for the quantity
greater than q(B).

In investigating the effect of changes in world
market prices on EC production, we have to work out
two different relationships: the effect of changes in
world market prices on marginal producer revenue and
the effect of changes in marginal producer revenue on
supply.

Changes in world market prices will not affect
revenue as long as producers supply no more than q(A).
In fact, they exploit the B quota almost completely.
This indicates that the marginal cost curve intersects
the individual price-sale function in the B quota
range. Hence, we have to investigate how the price
p(B) changes with variations in world market prices.

From Equation (31) we get for the rate of change
for p(B):

$$\frac{dp(B)}{p(B)} = \frac{p(w)}{p(B)} \cdot 0.6 \cdot \frac{dp(w)}{p(w)} \tag{32}$$

Equation (32) clearly shows that if $p(w) < p(B)$, the percentage change of the price $p(B)$, which is equal to marginal revenue, is always smaller than the percentage change of world market prices. If $p(w) > p(B)$, as it is temporarily, a 1 percent change in world market prices may lead to a change in $p(B)$ greater than 1 percent. This clearly shows that in times of already high world market prices, any change in these prices leads to an even greater change in the marginal revenue for EC sugar beet producers. The supply response of EC producers to changes in marginal revenue is not only a function of the EC supply curve but of the change in the production constraint as well. For example, if the domestic supply curve intersects the price-sale function in the vertical section between the B quota and the C quota, farmers will produce the quantity $q(B)$. The empirical evidence indicates that this is true for most EC sugar beet producers. If world market prices increase up to $p(B)$, the production constaint is less binding. The argument may be clarified with the help of Figure 2.

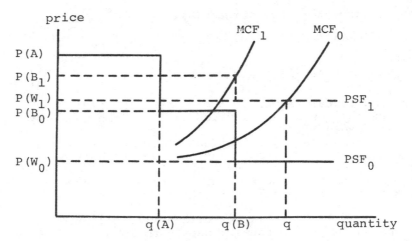

PSF   Price Sale Function
MCF   Marginal Cost Function

Figure 2   Supply Response of EC Sugar Producer Due to a Change in World Market Prices

We assume that quota B restricts production in the
base period. The relevant marginal revenue in this
period is p (B_0). If world market prices go up to
p(w_1), the price for quota B will increase to p(B_1).
The percentage change of p(B) is smaller than the
percentage change in world market prices. However, the
relevant marginal revenue is not p(B_0) any more, but
p(w_1). Hence, given the marginal cost curve MCF_0,
domestic production will expand up to q. A minor
change in marginal revenue leads to a higher response
in production than without the original binding
constraint. This is not true if the world market price
after the change is still below the initial domestic
price p(B_0). In this case, production will change only
with respect to the normal supply elasticity. This
shows that due to the institutional arrangements of the
EC sugar regime, the EC supply reaction to changes in
world market prices is distorted. Small changes in
world market prices may have no effect or only a minor
effect on EC supply. Large changes in world market
prices may lead to an overreaction of EC sugar
producers. This mechanism explains why the area
cultivated with sugar beets fluctuated more than the
area cultivated with sugar cane. Assuming that only
the EC consumer prices are fixed and EC producer prices
are allowed to very with world market prices, EC-
liberalization stabilizes the world market prices
irrespective of the price expectation procedure (see
Table 3). With static price expectations the
instability index decreases from 12.98 percent to 12.02
percent. We now have a plausible hypothesis and
evidence that EC sugar policy increased the amplitude
and length of the world market price cycle.

REFERENCES

Bale, M. D. and E. Lutz. "The Effectsof Trade
    Intervention on International Price Instability."
    61 (1979): 512-515.
Bale, M. D. and E. Lutz. "Price Distortions in
    Agriculture and Their Effects: An International
    Comparison." American Journal of Agricultural
    Economics." 63 (1981): 8-22.
Bartens/Mosolff (ed.). Sugar Economy 1981/82. Berlin,
    1982.
Cuddy, J.D. A. and P. A. Della Valle. "Measuring the
    Instability of Time Series Data." Oxford Bulletin
    of Economics and Statistics, 40 (1978): 79-85.
Deutscher Bundestag (ed.). Agrarbericht - Materialband.
    Bonn, Drucksache 9/2403, 1983.
FAO. Production Yearbook 1981. Rome, Food and
    Agricultural Organization

Gemmill, G.   The World Sugar Economy: An Econometric
    Analysis of Production and Policies.   Dissertation,
    Michigan State University, 1976.
Gemmill, G.   "Asymmetric Cobwebs and the International
    Supply of Cane-Sugar."   Journal of Agricultural
    Economics.   29 (1978): 9 - 21.
Grosskopf, W.   EG-Zuckermarktpolitik.   (Landwirtschaft-
    Angewandte Wissenschaft), Heft 224, Münster-
    Hiltrup, 1979.
International Sugar Organization.   London, Sugar Year
    Book 1980, 1981.
Johnson, D. G.   "World Agriculture, Commodity Policy,
    and Price Variability."   American Journal of
    Agricultural Economics.   57 (1975): 823-828.
Josling, T.   Developed-Country Agricultural Policies
    and Developing-Country Food Policies: The Case of
    Wheat.   "Research Report", No. 14, Standford,
    International Food Policy Research Institute, 1980.
Josling, T.   "Domestic Agricultural Price Policies and
    Their Interaction Through Trade."   McCalla, A.
    and T. Josling (ed.).   Imperfect Markets in
    Agricultural Trade.   Montclair, N.J., 1981.
Kirschke, D.   Trade Uncertainty in the Fixed Price
    Protective System.   Diskussionsbeiträge Nr. 49,
    Kiel, Institut für Agrarpolitik and Marktlehre,
    1983.
Koester, U.   Policy Options for the Grain Economy of
    the European Community: Implications for
    Developing Countries.   Research Report 35,
    Washington, International Food Policy Research
    Institute, 1982.
Koester, U. and P. M. Schmitz.   "The EC Sugar Market
    Policy and Developing Countries."   European Review
    of Agricultural Economics.   9 (1982): 183-204.
Lutz, E. and M. D. Bale.   "Agricultural Protectionism
    in Industrialized Countries and its Global
    Effects: A Survey of Issues."   Aussenwirtschaft
    (The Swiss Review of International Economic
    Relations).   5 (1980: 331-354).
Mood, A. M.;   F. A. Greybill and D. C. Boes.
    Introduction to the Theory of Statistics.   Third
    Edition, McGraw-Hill, Tokyo, 1974.
Ryll, E.   Der Einfluß agrarmarktpolitischer Instrumente
    der EG auf angebotsinduzierte
    Weltmarktspreisschwankungen unter Berücksichtigung
    additiv und multiplikativ verknüpfter
    Störvariablen. Diskussions-beiträge, Kiel, Nr.
    39, Institut für Agrarpolitik and Marktlehre, 1981.
Sampson, G. P. and R. H. Snape.   "Effects of the EEC's
    Variable Import Levies."   Journal of Political
    Economy.   88 (1980): 1026-1040.
Sarris, A. H.   "Commodity-Price Theory and Public
    Stabilization Stocks."   Chisholm, A. H. and R.

Tyers (ed.). Food Security: Theory, Policy, and Perspectives from Asia and the Pacific Rim. Lexington, Mass. and Toronto, 1982.

Schmidt, E. "Der Markt für Zucker." Agrarwitschaft, Jg 31, Heft 12 (1982): 393-399.

Schmitz, P. M. and U. Koester. "Der Einfluß der EG-Zuckermarkt-politik auf die Entwicklungsländer." Disskussionsbeiträge, Kiel, Nr. 42, Institut für Agrarpolitik and Marktlehre, 1981.

Schrader, J. V. "Interdependenzen zwischen EG-Zuckerpolitik and Preisoder Mengenschwankungen auf dem Weltmarkt." Agrarwirtschaft, Jf 31, Heft 1 (1982): 6-15.

Shei, S. Y. and R. L. Thompson. "The Impact of Trade Restrictions on Price Stability in the World Wheat Market." American Journal of Agricultural Economics. 59 (1977): 628-638.

Svedberg, P. "EEC Variable Import Levies and the Stability of International Grain Markets." Indian Journal of Agricultural Economics. 36 (1981): 58-66.

Tangermann, S. "Agricultural Trade Relations Between the EC and Temperate Food Exporting Countries." European Review of Agricultural Economics. 5 (1978): 201-220.

Tangermann, S. "Policies of the European Community and Agricultural Trade with Developing Countries." Johnson, G. and A. Maunder (ed.). Rural Change, Proceedings, Seventeenth International Conference of Agricultural Economists, Banff, Canada, 3rd-12th September, Farnborough. (1979): 440-453.

Turnovsky, S. J. "Stabilization." Journal of Public Economics. 9 (1978a): 37-57.

Turnovsky, S. J. "The Distribution of Welfare Gains from Price Stabilization: A Survey of Some Theoretical Issues." Adams, F. G. and S. A. Klein (eds.). Stabilizing World Commodity Markets. Lexington, Mass. and Toronto, (1978b); 119-148.

Zwart, A. C. and K. D. Meilke. "The Influence of Domestic Pricing Policies and Buffer Stocks on Price Stability in the World Wheat Market." American Journal of Agricultural Economics. 61 (1979): 434-445.

APPENDIX

First-order and second-order derivatives of the world price function with respect to the random supply disturbances θ and ϕ

a) Free Trade Policy: $d(p,\theta)+f(p,\phi)-g(p)-h(p) = 0$

$$p_\theta = \frac{-d_\theta}{d_p+f_p-g_p-h_p} < 0, \quad p_\phi = \frac{-f_\phi}{d_p+f_p-g_p-h_p} < 0$$

$$p_{\theta\theta} = \frac{2d_{\theta p}d_\theta[d_p+f_p-g_p-h_p]-[d_{pp}+f_{pp}-g_{pp}-h_{pp}]d_\theta^2}{[d_p+f_p-g_p-h_p]^3}$$

$$p_{\phi\phi} = \frac{2f_{\phi p}f_\phi[d_p+f_p-g_p-h_p]-[d_{pp}+f_{pp}-g_{pp}-h_{pp}]f_\phi^2}{[d_p+f_p-g_p-h_p]^3}$$

$$p_{\theta\phi} = \frac{[d_{\theta p}f_\phi+f_{\phi p}d_\theta]\cdot[d_p+f_p-g_p-h_p]-[d_{pp}+f_{pp}-g_{pp}-h_{pp}]d_\theta}{(d_p+f_p-g_p-h_p)^3}$$

b) EC Price Fixing Policy: $d(\bar{p},\theta)+f(p,\phi)-g(\bar{p})-h(p) = 0$

$$p_\theta = \frac{-d_\theta}{f_p-h_p} < 0, \quad p_\phi = \frac{-f_\phi}{f_p-h_p} < 0$$

$$p_{\theta\theta} = \frac{[f_{pp}-h_{pp}]d^2}{[f_p-h_p]^3}$$

$$p_{\phi\phi} = \frac{2f_{\phi p}f_\phi[f_p-h_p]-[f_{pp}-h_{pp}]f_\phi^2}{[f_p-h_p]^3}$$

$$p_{\theta\phi} = \frac{f_{\phi p}d_\theta[f_p-h_p]-[f_{pp}-h_{pp}]d_\theta f_\phi}{[f_p-h_p]^3}$$

TABLE A1
Data for Quantifying the Effect of the EC Trade
Liberalization on World Market Price Stability.

	European Community (9)	Rest of the World
Production[1]	12,279,371 t	65,361,666 t
Consumption[1]	8,384,973 t	69,256,064 t
Net Export[1]	3,894,398 t	- 3,894,398 t
Production Share	15.8%	84.2%
Trade Share	21.7%	78.3%
Producer Price[2]	314.79 US-$/t	155.76 US-$/t
Consumer Price[3]	503.67 US-$/t	155.76 US-$/t
Supply Elasticity[4]	0.3(1.5)	0.4(1.0)
Demand Elasticity[4]	-0.1(-0.3)	-0.5(-1.0)

[1] Tons of white sugar 1980. Raw sugar data were converted into white sugar using a conversion factor of 0.92.

[2] EC intervention price for white sugar of the maximum quota 1982/83. Daily raw sugar price (june 1982) of the International Sugar Agreement for the rest of the world adjusted by the conversion factor of 0.92.

[3] EC intervention price for white sugar of the basic quota 1982/83.

[4] Short and long-term elasticities.

Sources: International Sugar Organization (ed.), Sugar Year Book 1980. London 1981. Koester and Schmitz, The EC Sugar Market Policy and Developing Countries. "European Review of Agricultural Economics", Vol. 9, No. 2 (1982), p. 202. Schmidt, Der Markt für Zucker. "Agrarwirtschaft", Jg. 31, Heft 12 (1982), pp. 393-399. Deutscher Bundestag (ed.), Agrarbericht-Material-Band. Drucksache 9/2403, Bonn 1983, p. 114.

11
Commodity Price Stabilization in Imperfectly Competitive Markets

David M. Newbery

INTRODUCTION

In Newbery and Stiglitz (1981), we developed a
reasonably comprehensive analysis of commodity price
stabilization on the assumption that both producers and
consumers were price takers. We recognized that trade
in many commodities is subject to a variety of
distortionary interventions, such as tariffs, quotas,
domestic subsidies, acreage restrictions and the like,
and analysed the impact of international commodity price
stabilization in a distorted world economy (Newbery and
Stiglitz, 1981, Ch. 19, pp. 272-83). However, our
analysis of distorted markets assumed that the structure
of protection and distortion would not be affected by
international price stabilization, whilst individual
producers and consumers remained price takers within
this distorted framework. In short, governments were
not assumed to be intervening to exercise international
market power, and hence would have no reason to change
the pattern of intervention in response to changes in
the world market.

Clearly it would be desirable to try and extend the
analysis of commodity price stabilization to deal with
imperfectly competitive commodity markets. At least
some countries have a sufficiently high share of world
trade in specific commodities to give them some market
power.

The popular example today would no doubt be OPEC
oil, though it is difficult to determine how far OPEC's
market power is exercised by a well coordinated cartel,
or subcoalition, or how far it depends on the actions
of the largest member, Saudi Arabia. Before 1973, the
examples which might have sprung to mind were Brazilian

David M. Newbery, World Bank, Washington, D. C.

coffee or Indian tea. Table 1 lists countries which

TABLE 1
Shares of Individual Countries in World Trade of
Commodities, Averages 1977-79

Country	Product	Percentage
>50 percent		
USA	Maize	73
Philippines	Coconut Oil	71
Bangladesh	Jute (c)	70
Malaysia	Palm Oil	70
Argentina	Linseed Oil	69
Australia	Wool (greasy)	60
Malaysia	Copra	55
Malaysia	Rubber (c)	51
25-50 percent		
USA	All Cereals	49
Brazil	Sisal	42
USA	Wheat	39
Malaysia	Tin (c)	36
Morocco	Phosphate Rock	34
Senegal	Groundnut Oil	29
USA	Cotton (c)	29
Tanzania	Sisal (c)	28
India	Tea (c)	27
Saudi Arabia	Oil	26
Guinea	Bauxite	25
Canada	Barley	25
Cuba	Sugar (c)	25

(c) = Core commodity.

Sources: Commodity Trade and Price Trends, World Bank,
1981; FAO Trade Yearbook, 1979.

had more than 25 percent of world trade in a given
commodity on average over 1977-79. Interestingly,
Brazilian coffee does not appear (only 17 percent),
whilst Saudi oil is 26 percent, near the cut-off point
on the list. Of course, trade shares in specific
commodities are an imperfect measure of market power,
which depends on elasticities of demand and rest of
world supply, as well as the extent of domestic
production and transport costs, both of which define
the size of the relevant market. Amongst the vegetable

oils substitution possibilities will clearly limit the exercise of market power, whilst synthetic substitutes likewise constrain the fibre and rubber markets. On the other hand, transport costs may make regional markets more relevant than the whole world market for some commodities whilst a given country's commodity may be sufficiently differentiated from grades supplied by other countries that the country can exercise some market power despite its low overall market share. Nevertheless, despite all these qualifications, it seems reasonable to conclude that some countries (or cartels of countries) have some market power in some commodity markets, and that it is therefore worth enquiring how this market power affects the issue of international price stabilization.

The paper is organised as follows. Section 2 summarises the analysis of the case for international price stabilization in distorted markets referred to above in the first paragraph. Most distortions are probably not designed to exercise market power, yet clearly have an impact on the market, and hence on the case for stabilization. The remainder of the paper studies the case in which a producer exercises deliberate market power. In section 3 the various questions this raises are posed, and then answered in the remaining sections. Section 4 computes the optimum stocking rule for a monopolist and compares it with the competitive rule, asking whether he will undertake more or less price stabilization, and on what this depends, and what it implies for overall price stability, given the actions of the other agents. Section 5 investigates the impact of an international buffer stock agency on an imperfectly competitive market.

DO MARKET IMPERFECTIONS STRENGTHEN THE CASE FOR COMMODITY PRICE STABILIZATION?

In some commodities, notably foodstuffs, international trade is subject to a variety of distortionary interventions. (See, e.g. Bale and Lutz, 1981; Lutz and Scandizzo, 1980.) These typically emerge when local producers find themselves in competition with imports, and are quantitatively important for sugar, cereals, rice, and to a lesser extent, meat and dairy products. In most cases, it is difficult to argue that these policies benefit the importing country as a whole, for, with the notable exception of the U.S., the countries imposing the restrictions are small relative to world trade in the commodity. (Arguably, the EEC is the other main exception, though it seems unlikely that the main motive for the Common Agricultural Policy was to improve the EEC's terms of trade, though this may well be one of its consequences.) If this analysis is accepted (for some commodities, and some countries) then the market distortions are not evidence of market power, and it is possible to retain the simplifying assumption that producers and consumers are price takers, though the prices they face will no longer be uniform throughout the market. Much of the conventional analytics of price stabilization theory will still apply, though the distortions introduce additional complications, especially for the welfare analysis.

Newbery and Stiglitz (1981, Ch. 19, pp. 272-83) have already analyzed this case, and their results are briefly summarized here for completeness. They asked whether the presence of trade distortions increased or reduced the amount of price stabilization which was desirable compared to the level which would be desirable in the absence of distortions. They argued that price stabilization may generate larger benefits in the presence of trade distortions than in their absence for at least three quite different reasons:

(1) It is often argued that world and domestic prices are more volatile in the presence of trade restrictions than in their absence, and less use is made of the risk-sharing aspects of international trade. If stabilization is desirable under free trade then the greater instability under restricted trade makes it even more desirable. Between 1972 and 1974 the U.S. wholesale grain price more than tripled, while in the EEC it rose by 20 percent, and in the USSR it apparently remained unchanged. Johnson (1975) argues that the international market could have absorbed the modest world production shortfalls with only modest price increases, but because the shortfalls were only shared by part of the world market the price instability was severe.

At this point, it should be stressed that the link between trade restrictions, market fragmentation, and price volatility is not simple, and depends on the kind of trade policy adopted. We built a simple model to explore this issue, in which we ignored supply responses, distributional issues, and attitudes to risk (all of which might be important) and compared the benefits of price stabilization with and without distortions.

We found that _linear_ uncorrelated trade policies (in which there is a linear relationship between domestic and world prices, as with tariffs, provided the tariff wedge is not systematically correlated with the size of disturbances) do not affect the benefits of price stabilization significantly, but _non-linear_ policies do. In particular, if countries use quotas rather than tariffs, then even if they do not change with price stabilization, there may be additional benefits over and above the benefits conventionally

calculated (assuming no distortions). The reason is
that with linear policies the average degree of
distortion does not change as the variability of price
changes, but with non-linear policies it may be reduced.
    (2) If the total world consumption of commodities
subject to distortion is increased, then there is an
additional gain (equal to a reduction in the excess
burden of the distortion) equal (roughly) to the
increase in supply multiplied by the difference between
consumer and producer price. This benefit is an
important component in cost-benefit analysis (see e.g.
Boadway, 1974, or Harberger, 1971). The argument is
then that price stabilization induces a positive supply
response which then generates extra efficiency gains
(over and above those realized on competitive markets).
Whether in fact price stabilization will lead to
increased supply is discussed in Newbery and Stiglitz
(1981, Part V) where it will be seen that the answer is
by no means obvious.
    (3) It has been cogently argued (FAO, 1975) that
if price stabilization is achieved by creating buffer
stocks, then this will provide an impetus to highly
desirable trade liberalization. The argument here is
that restricted trade and the resulting price
volatility make both the availability and cost of
supplies so uncertain to importing countries that they
create farm support programmes to ensure supplies, and
impose further restrictions to make them viable. If
the world price were stabilized and supplies guaranteed,
then these restrictive measures would be unnecessary
and might be dismantled. Even where the original
intervention was inefficient its true cost may remain
obscure until the price is stabilized. Once the cost
is evident, the source of the inefficiency might be
removed.
    This last argument may well be the most important,
but it requires a theory of the choice of trade policy
before it is possible to say how institutional change
may affect the structure of trade distortion. Newbery
and Stiglitz (1981, Chapter 24), go some way towards a
theory of the choice of trade policy under risk, but they
ignore the political economic aspects of agricultural
price intervention, which are also clearly important.[1]

PRICE STABILIZATION IN A MONOPOLIZED COMMODITY MARKET

    In a competitive market economy with a complete
set of futures and insurance markets the market
equilibrium is Pareto efficient and market intervention
by, for example, an International Buffer Stock Agency,
to further stabilise commodity prices, would be
inefficient. In such a market, private agents would
provide the efficient level of storage activity and

price stabilization. Now, whilst it can frequently be argued that commodity markets are competitive, it is clear that the market structure is not complete, lacking insurance markets and most futures markets. In such cases Newbery and Stiglitz (1982a) show that in general a competitive market equilibrium is not even constrained Pareto efficient, so that in principle a Government could make everyone better off by market intervention (e.g. setting taxes, or subsidising storage activities).

However, they also show that if agents are risk-neutral and hold rational expectations, then the competitive equilibrium will be efficient. The reason is obvious. If agents are risk-neutral, and hold common (objective) beliefs about the economy, then they would not wish to trade on risk markets even if they existed. Similarly, if they hold common (objective) beliefs about futures prices, they would not wish to trade on futures markets, and in both cases such markets would be redundant. Their absence therefore makes no difference to the market equilibrium, which will remain Pareto efficient.

It will greatly simplify the analysis if we continue to make these assumptions of risk-neutrality and rational expectations, though a full analysis would clearly relax these assumptions. (See, in particular, Newbery and Stiglitz, 1981, Chapters 13, 14.1, 15 and 30; and Newbery and Stiglitz, 1982b.) In making these assumptions we are following tradition (e.g. Gustafson, 1958; Samuelson, 1971), though most writers appear unaware of the strength of these often implicit assumptions. Under certain additional assumptions (stable linear demand, stationary stochastic supply, and some restrictions on various parameters) it is possible to solve analytically the competitive (and thus efficient) stockpiling rule (see Newbery and Stiglitz, 1981, Ch. 30; 1982b). At this stage the main point to notice is that the rule is very different from the favored intervention rule proposed for International Buffer Stock Schemes, which would attempt to keep prices within a specified band width. The efficient rule would store a fraction (somewhat greater than one half) of the excess of harvest plus last year's carryover above some critical amount. (See below.) This rule remains a good approximation (and a natural starting point for iterative numerical solution) when the simplifying assumptions do not hold. (See Newbery and Stiglitz, 1981, Ch. 30; Gustafson, 1958.)

Given this benchmark of the competitive stocking rule for the efficient level of price stabilization, a number of interesting questions can now be asked of an imperfectly competitive commodity market. Again, in the interests of simplicity, we shall confine attention to a monopolist facing a large number of competitive consumers. First, on the assumption that the producer does the storage and price arbitrage, how does the amount of storage and hence price stabilization compare under monopoly with the efficient level? Do monopolists perform more or less price stabilization? The answer to this question immediately raises related questions - under what circumstances do monopolists do less stabilization, and which agents, consumers or producers, will carry the stock? Again, the answer will affect the next question, which is, if consumers have a comparative advantage in carrying the stocks, how does the change in their demands (for consumption and storage) affect the market power of the monopolist?

The next set of questions concern the consequences of establishing an International Buffer Stock to stabilize prices. One might first ask what effect this would have on the market if the Buffer Stock Agency followed conventional rules, and did not attempt to use its potential market power. Do different buffer stock rules (e.g. band width rules or competitive rules) differ in their robustness to monopoly manipulation?

Finally, what might happen if the Buffer Stock Agency were able to use its market power to countervail the monopolist? (The more plausible scenario, in which the Agency acts as a cartellizing influence on the primary producers is subsumed in the analysis which treats the monopolist as stockholder.)

THE OPTIMAL STOCK RULE FOR A MONOPOLIST

The simplest model which allows an analytical solution in the competitive case (see Newbery and Stiglitz 1982b) contains the following elements:
    (1) There is a stock S_{t-1} carried forward from the previous year.
    (2) To this is added a random harvest, $\tilde{h}_t$, so that at date t the amount available for consumption, C_t, and for storage, S_t, is the total supply, x_t:

$$x_t = h_t + S_{t-1} = C_t + S_t. \tag{1}$$

We assume that there are no losses in storage (though these are easy to handle - see Samuelson, 1971), and that planned production does not change each year[2] (e.g., for a tree crop like coffee). If weather and other random factors are serially uncorrelated, this implies that the harvest is also a serially uncorrelated, stationary random variable. We choose units so that the average harvest is one unit:

$$E\tilde{h}_t = 1, \ \text{Var } \tilde{h}_t = \sigma^2, \ \text{Cov}(h_t, h_{t'}) = 0, \ t \neq t' \tag{2}$$

Finally, we assume that demand is stationary and

non-stochastic, so that the market clearing price p_t, depends only on current consumption, C_t:

$$p_t = p (C_t).\tag{3}$$

When it comes to solving explicitly for the storage rule we shall assume that the demand schedule is linear, with elasticity ε at the prestabilization mean price, $\bar{p}$. If, on average, consumption equals harvest, so that stocks neither continually increase nor decrease, the (inverse) demand schedule will be

$$p(C) = \bar{p}\{1 - \frac{1}{\varepsilon} (C-1)\}.\tag{4}$$

Social welfare can be measured in money units, since we assume risk neutrality and ignore income distribution, as

$$U(C) = \bar{p}\{(1 + \frac{1}{\varepsilon})C - \frac{1}{2\varepsilon}C^2\},\tag{5}$$

so that

$$p(C) = \frac{dU}{dC}.$$

The competitive storage rule can be found by maximising expected social welfare (the approach taken by Gustafson, 1958, and Samuelson, 1971) or derived directly from the competitive arbitrage conditions. It is easy to demonstrate the equivalence of these two approaches (see Newbery and Stiglitz, 1982b) so we shall follow the second, more transparent approach.

If the annual storage costs excluding interest is k per unit at the margin, then a speculator who buys after the harvest at price p_t and sells after the next harvest at price p_{t+1} will have made a marginal profit (in money terms at date t + 1) of

$$p_{t+1} - (p_t + k)(1 + r)$$

per unit, where r is the rate of interest. If speculators are risk neutral then they will store nothing if

$$E\ p_{t+1} < (p_t + k)(1 + r),$$

but will otherwise continue to store until they have driven up the current price and driven down the expected future price to the point where

$$(p_t + k)(1 + r) = E\ p_{t+1}.$$

These two cases can be combined in the fundamental competitive arbitrage equation

$$p_t + k \stackrel{\geq}{=} \beta\ E\ p_{t+1} \left.\vphantom{\begin{matrix}a\\b\\c\end{matrix}}\right\} \qquad \text{complimentarily} \quad (6)$$

$$S_t \stackrel{\geq}{=} 0$$

where $\beta = 1/(1 + r)$ is the discount factor. Our first objective is to investigate the form of the competitive storage rule, and this appears to be difficult because the expected price next year depends on planned carryovers in the following year. To find the storage rule now we need to know the storage rules to be followed hereafter. To solve the problem we need to start with a terminal date at which the carryforward is specified, and hence known, and work backwards. This is the standard method of solving stochastic programming problems, but with a long time horizon is computationally demanding. If, however, the time horizon is allowed to lengthen and if the expected present discounted value of terminal stocks, $\beta^T S_T EP_T$, tends to zero, then the influence of the future on the present also tends to zero, and we can instead of looking for a particular solution (which depends on S_T), look for a stationary stock rule which is independent of terminal stocks, and hence independent of time. Since price only depends on consumption from (3) and since from (1)

$$C_t = x_t - S_t \qquad (7)$$

it follows that we are looking for a stock rule which is a function only of total supply x_t:

$$S_t = f(x_t) \stackrel{\geq}{=} 0 \qquad (8)$$

Contrast the form of this rule with the bandwidth rule, where if the demand is $C = D(p)$ and the price band has upper and lower intervention points p^u, p^f, then the stock rule is

$$S_t = f(x_t, h_t) = \begin{cases} x_t - D(p^f) & h_t > D(p^f) \\ x_t - h_t = S_{t-1} & D(p^u) \stackrel{\leq}{=} h_t \stackrel{\leq}{=} D(p^f) \\ x_t - D(p^u) & h_t < D(p^u)\ , \quad (9) \\ & x_t > D(p^u), \\ 0, & \text{otherwise} \end{cases}$$

The bandwidth rule has storage depending on both supply

and harvest over the central range, and thus is not a competitive arbitrage rule. More importantly, it follows that it is not an efficient method of stabilising consumption in a competitive market, which, given our assumptions about risk neutrality, is the basic reason for stabilising prices.

To return to the problem of finding the competitive stock rule, we seek a function $f(x)$ which solves equation (6), which, given (3), (7) and (8) can be written

$$p\{x_t - f(x_t)\} + k \overset{\geq}{=} \beta \; Ep \left[\tilde{h}_{t+1} + f(x_t) \right.$$

$$\left. - f\{\tilde{h}_{t+1} + f(x_t)\} \right] \quad \text{and} \; f(x_t) \overset{\geq}{=} 0 \quad \begin{array}{c} \text{compliment-} \\ \text{arily} \end{array} \quad (10)$$

The problem in solving this arises from the righthand side, in the term $f\{h + f(x)\}$. One key feature of the rule is, however, immediate.

The optimum storage rule is non-linear. This follows because stocks must be non-negative, so $f(x) = 0$ below some critical value of x_o, for which

$$p(x_o) + k = \beta \; E \; p(\tilde{h} - f(\tilde{h})). \quad (11)$$

For the linear demand function of equation (4) equation (10) can be rearranged to yield

$$f(x) \quad \begin{array}{l} = \dfrac{1}{1 + \beta} \; [x - a + \beta \; Ef\{\tilde{h} + f(x)\}], \quad x \geq x_o, \\[2mm] = 0, \hspace{5cm} x < x_o, \end{array} \quad (12)$$

where x_o is defined by (11) and $\underline{a}$ is a constant:

$$a = 1 + \varepsilon(1 - \beta + k/\overline{p}) \overset{\sim}{=} 1 + \varepsilon(c + r), \quad c \equiv k/\overline{p}. \; (13)$$

Here c is the total annual storage cost per \$ stored excluding interest and c + r is the total cost including interest. In Newbery and Stiglitz (1982b) we show how to solve equation (12) for the special case of the two point distribution

$$h \quad \begin{array}{l} = 1 + u \quad \text{with Prob} \quad \rho \\[2mm] = 1 - \gamma u \; \text{with Prob} \; 1 - \rho \end{array} \quad \gamma = \dfrac{\rho}{1-\rho}, \quad (14)$$

so that

$$E\tilde{h} = 1, \quad \text{Var} \; \tilde{h} = \gamma u^2 \equiv \sigma^2.$$

In this case the competitive stocking rule is linear beyond x_o:

$$f(x) = \alpha(x - x_o), \qquad x_o \lessapprox x \le x_m, \tag{15}$$

where

$$\alpha = \frac{1 + \beta - \sqrt{(1 + \beta^2) - 4\beta\rho}}{2\beta\rho} \tag{16}$$

$$x_o = \frac{a - \alpha\beta\rho(1+u)}{1 - \alpha\beta\rho}, \quad x_m \equiv \frac{1+u - \alpha x_o}{1 - \alpha}. \tag{17}$$

(x_m is the maximum conceivable value for x) provided the parameters lie in a certain range:

$$1 - \gamma u + \alpha(x_m - x_o) < x_o < 1 + u. \tag{18}$$

That this is the solution to (12) can be checked by substitution. The intuitive reason for both the linearity (beyond x_o) and the constraints of (18) is that it must be the case that if the harvest is high $(1 + u)$ then positive stocking is undertaken, even if current stocks are zero, whilst if the harvest is low $(1 - \gamma u)$ then there is no carry forward, even if current stocks are high (e.g. at their maximum value, $S_m = \alpha(x_m - x_o)$). Then the current state of the system is essentially determined by the current harvest, which can only take two values. Hence, to check that the parameters are such that equation (15) is the solution, once β, ε, $k/\bar{p}$, u, and ρ have been specified, α, x_o and x_m can be computed from (16) and (17) and checked to see if (18) is satisfied.

So far we have derived the competitive rule, but fortunately the monopoly stocking rule now follows immediately, for the monopolist is interested in arbitraging marginal revenue, rather than price. Since for a linear demand schedule the marginal revenue is also linear, the dominant producer's problem is isomorphic to the competitive problem. Instead of the arbitrage rule of equation (6) we have

$$m_t + k \gtrapprox \beta Em_{t+1}$$
$$S_t \ge 0 \qquad \} \text{ complimentarily} \tag{19}$$

where m_t is the marginal revenue at date t. For the linear demand schedule of equation (4)

$$m(C) = \bar{p}\left(1 + \frac{1}{\varepsilon} - \frac{2C}{\varepsilon}\right), \quad C = x - f(x). \tag{20}$$

Equation (19) and (20) give rise to a storage rule which is almost identical to that of equation (12), except for the constant term a:

$$f(x) = \frac{1}{1+\beta} [x - A + \beta Ef \{\tilde{h} + f(x)\}] \ , \ x > x^m_o \ ,$$

(21)

$$A = \frac{1}{2} (a + \beta), \ x^m_o = \frac{A - \alpha\beta\rho(1 + u)}{1 - \alpha\beta\rho}$$

which can be solved to give

$$f(x) = \alpha(x - x^m_o), \ x \geq x^m_o.$$

(22)

The value of α is the same as before, and given by (16), whilst the intercept, x^m_o, is similar in form to (17), except a is replaced by A. Thus the monopoly stocking rule has the same slope, or marginal storage rate, α, but leads to an unambiguously larger level of storage, since $x^m_o < x_o$ because $\beta < 1 < a$. Figure 1 illustrates.

stock $f(x)$

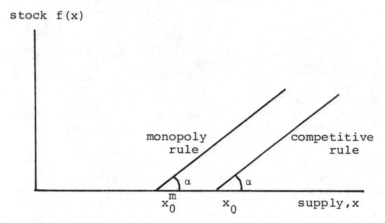

Figure 1 Monopoly and Competitive Rules Compared

Numerical Example

If $\rho = 0.75$ (Bad harvests once every four years)
        $u = 0.20$ (Harvests 1.2 or 0.4, coefficient of
                    variation (CV) of output 35 percent)
        $\varepsilon = 2.5$ (elasticity, unstabilized CV of
                    price = 13.9 percent)
        $\beta = 0.95$ (5 percent interest rate)
      $k/\bar{p} = 3$ percent (storage costs)
then a = 1.2, $\alpha = 0.6835$, and the competitive storage
rule is
$$f^C(x) = 0.6835 (x - 1.2), \ x \geq 1.2$$

(23)

(Thus, since harvests never exceed 1.2, there is never any storage in the competitive case, and prices would not be stabilized any further.) The monopoly storage rule is

$$f^m(x) = 0.6835 \ (x - 0.9563) \ , \ x \geq 0.9563.$$
(Maximum supply $x_m = 1.7263$, maximum stockpile 0.5263.) (24)

This stocking rule implies average stocks of 25.6 percent of average harvest, which is substantial. Moreover, it reduces the CV of consumption from 35 percent to 22 percent, and lowers the CV of prices from 14 percent to 9 percent.

The effect of stockpiling in this example is to substantially improve monopoly profits and reduce average consumer surplus relative to the no stabilization monopoly equilibrium. The extra loss of consumer surplus averages about 9 percent of (riskless) consumer expenditure on the crop. The result of this section can now be summarized as follows.

Proposition 1. A monopolist facing a stable linear demand schedule will undertake more price stabilization through storage activities than a competitive market producing the same average supply, and will thereby be able to exploit the consumer more effectively.

This immediately raises the next question.

When do monopolists stabilize more than the competitive market?

Consider a situation in which, with a given level of current supply, it is just not worth a competitive market storing, so that

$$p(Q_0) + k = \beta E p(Q_1)$$
(25)

where period 0 is today, and period 1 is the next period. In such cases it will pay a monopolist to store if

$$\frac{dR(Q_0)}{dQ_0} + k < \frac{\beta E \ dR(Q_1)}{dQ_1}$$
(26)

where $R(Q)$ is revenue from the sale of Q. This can be rewritten as

$$p(Q_0)(1 - \frac{1}{\varepsilon_0}) + k < \beta E p(Q_1)(1 - \frac{1}{\varepsilon_1})$$

where $\varepsilon_i = \varepsilon(Q_i)$ is the elasticity of demand when sales are Q_i. Substituting from (25), the monopolist will store more if

$$\varepsilon o/_{\overline{\varepsilon}} < 1 - (1 + r)k/_{\overline{p}} \qquad (27)$$

where r is the rate of interest and $\overline{\varepsilon}$ is an average price elasticity:

$$\overline{\varepsilon} \equiv \frac{Ep}{Ep/\varepsilon} \; .$$

The accumulated proportional storage cost, $(1+r)k/_{\overline{p}}$, takes values between 1.3 percent and 5.1 percent for the six core commodities considered by Newbery and Stiglitz (1981, Table 20.7, p. 295). Equation (27) demonstrates immediately that if the demand schedule has constant elasticity the monopolist will do less stabilization than a competitive market, but this result is very sensitive to the shape of the demand schedule. A slight fall in price elasticity at lower prices (by, e.g. 5-10 percent) will induce the monopolist to undertake more stabilization. With a linear demand schedule (for which the elasticity falls rapidly) this tendency can be quite pronounced, as the previous section demonstrated.

Of course, there may be quite different reasons why a monopolist might prefer price stability. He might fear that temporarily high prices would induce new entrants into the industry and reduce his future profits, and hence might limit price (or, more to the point, maintain the equivalent of excess capacity in the form of buffer stocks to deter entry. See Spence, 1977). Ignoring such strategic issues, though, we can summarise these results as follows.

Proposition 2. A monopolist facing a stable constant elasticity of demand schedule or a schedule whose price elasticity falls as price rises will undertake less price stabilization than a competitive market producing the same average supply. Whether or not the monopolist does more or less stabilization depends sensitively on the shape of the demand schedule.

If monopolists undertake less stabilization than the competitive market, then we have to consider the possibility that other agents (consumers, or, more probably, independent stock holders) will perform further arbitrage, which will in turn alter the gross demand schedule (i.e. including demands for additions to storage) facing the monopolist. This may lead the monopolist to change his behavior. Thus it is important to ask:

Who performs the storage activities in an imperfectly competitive market?

A large number of factors are relevant even in the simpler competitive case, and some of them are reviewed in Newbery and Stiglitz (1981, Ch. 14, pp. 195-6). If

agents are not risk neutral, then stocks provide a
convenience yield in the form of insurance, and the
agent whose net cost of storage (carrying costs less
convenience yield) are lowest will have a comparative
advantage in storage. Thus, if the aggregate demand
facing producers is elastic, and prices are negatively
correlated with output (e.g. if supply shocks generate
price instability) then producers derive a positive
convenience yield for stocks. If, on the other hand,
demand fluctuations are the source of the price
instability, consumers will enjoy a positive convenience
yield unless prices vary with their incomes.
Intermediate producers (i.e. agricultural processors)
enjoy a positive convenience yield since their profits
will be negatively correlated with input prices.

On the other hand, transport and handling costs
will convey a cost advantage on producers, since
downstream users will have to incur additional interest
costs on the marketing margin. However, economies of
scale in storage, access to cheaper finance, hedging
facilities, better market information, and risk pooling
advantages may offset these cost disadvantages and
confer the comparative advantage on middle men. The
following section assumes that all these various factors
cancel out, so that producers, consumers and middle men
face equal net storage costs.

Consider the case in which the consumption demand
schedule is stable and has constant elasticity of demand.
The argument of the previous section then shows that, if
only producers stockpile, then a monopolist would
perform less stabilizing action than a competitive
market. Consequently, there would be enough remaining
price variability to justify further price stabilization
by competitive agents, assuming they face no cost
disadvantage. What would happen if such agents
intervened and collectively achieved a competitive
degree of price stabilization?

The analysis of Newbery and Stiglitz (1981, Ch. 30)
and Gustafson (1958) shows that the qualitative form of
the stocking rule remains unchanged as the shape of the
demand schedule and the form of the probability density
of harvests varies. The level of supply below which
there is no stocking (x_0) does depend on these features,
but can be found relatively easily. The stocking rule
will typically be a convex function of supply, whose
general shape is readily approximated by a piecewise
linear form. (Newbery and Stiglitz, 1981, pp. 435-438.)
If competitive consumers undertake the storage activity,
then the producers will face a displaced gross demand
schedule (for consumption plus storage additions) as
shown in Figure 2.

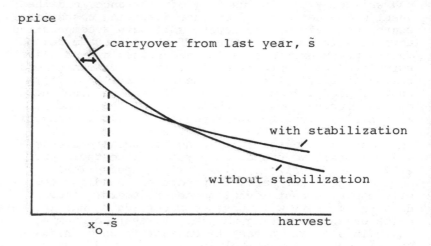

price

carryover from last year, $\tilde{s}$

with stabilization

without stabilization

$x_o - \tilde{s}$ harvest

Figure 2   Effective Demand Schedule Facing Monopolist

The form of the demand schedule is approximately

$$C = q + S = p^{-\varepsilon} \qquad\qquad q \overset{<}{=} x_o - S$$

$$C = \alpha x_o + (1 - \alpha)(q + S)\, p^{-\varepsilon} \qquad q \overset{>}{=} x_o - S \qquad (28)$$

where S is the carryover from last year, and α, x_o, represent the linear approximation to the competitive stocking rules of equation (15). The net effect of consumer stockpiling is to make the elasticity of demand <u>increase</u> as the supply increases and price falls, and hence to further discourage the monopolist from undertaking any storage. Thus, providing there is no cost disadvantage facing the consumer, they will presumably displace the monopolist and all storage will be undertaken by consumers.

Summarising this argument we have, depending on who does the stabilization:

<u>Proposition 3</u>. If consumers face the same or lower net storage costs (including interest charges) than producers, then prices should be either more stable, or at worst, no less stable, than in competitive markets handling the same average production.

If, on the other hand, consumers face cost disadvantages, then prices may be less stable under monopoly.

THE EFFECT OF INTERNATIONAL BUFFER STOCKS ON IMPERFECTLY
COMPETITIVE MARKETS

So far we have restricted attention to either the
producer or the competitive consumers, either of which
might carry stocks and hence stabilize prices. If,
however, an International Stabilization Authority is
set up, then it will presumably be given rules for
operating its intervention activities, and these rules
may affect the market facing the monopolist. It is
interesting to consider three cases. The first would
enquire into the likely consequences of introducing the
favored Bandwidth rule into the imperfectly competitive
market. The second case would examine the consequences
of introducing the buffer stock rule which would be
optimal in a competitive market, but which would not
necessarily be best for an imperfectly competitive
market. Finally, the most difficult but most interesting
question is to ask what rule would be optimal in an
imperfectly competitive market.

Bandwidth Rule

This rule is specified in equation (9) and
generates an effective demand facing the producer as
shown in Figure 3 below (assuming sufficient stocks are
on hand, i.e. at least h_1).

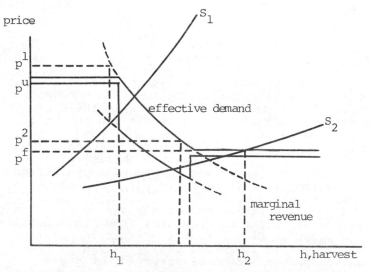

Figure 3   Effective Demand Under the Bandwith
           Rule

The same figure shows the marginal revenue facing the monopolist, and it is this schedule which is relevant for his production and storage decisions. First, note that if the monopolist retains any short run supply discretion once the stochastic uncertainty has been resolved, then the Bandwidth rule can actually destabilize prices. For suppose that in bad years the supply schedule is S_i, whilst in good years it is S_2. Without stabilization the price would be either p_1, or p_2, but with stabilization the monopolist would choose to supply h_1, or h_2 (significantly more variable) at prices p^u, p^f. It is quite likely in this case that the buffer agency will overaccumulate stocks since the floor price p^f may generate a very attractive marginal revenue for the monopolist.

If the producer has negligible short run supply discretion (i.e. the case considered in the previous section) then the appropriate question to ask of the international buffer scheme is whether it is sustainable, or whether it will induce speculative attacks which render it non-viable. The issue of vulnerability to speculative attack has been addressed by Salant (1979) and Newbery and Stiglitz (1981, p. 413-4). In a competitive market in which the upper and lower price bands are set at $\bar{p}(1+b)$, $\bar{p}(1-b)$, it will pay producers or speculators to speculate on a price rise whenever the price falls to the floor price unless

$$\bar{p}(1-b) + k > \beta Ep = \beta\bar{p} \tag{29}$$

or

$$b < 1 - \beta + k/\bar{p} \approx r + c , \quad c \equiv k/\bar{p} .$$

If the bandwidth is set further apart than this, then private agents will speculate against the stockpile. Moreover, even if the buffer agency sets the bandwidth appropriately, then if its own stocks are insufficient to sustain the ceiling price, the right hand side of (29) will be increased, since the expected next period price will be biased up by the probability of a shortfall too large to be met by the agency's buffer stock, resulting in a price above $\bar{p}(1 + b)$ and hence raising Ep above $\bar{p}$.

In a non-competitive market the agency is subject to additional speculative pressures from the monopolist, for it is clear in Fig. 4 that the expected marginal revenue next period is substantially raised by the intervention prices. Even if it would not pay a competitive speculator to buy now in the expectation of future profit, it might well pay a monopolist. As an example, consider the linear demand schedule of equation (12), and suppose the distribution of harvest

$\tilde{h}$ is symmetric about $\bar{h} = 1$. Let

$$\text{Prob } \{h \gtreqless 1 + \varepsilon b\} = \frac{1}{2} (1 - \Pi)$$

so that

$$\text{Prob } \{1 - \varepsilon b < h < 1 + \varepsilon b\} = \Pi$$

then, with no producer carry forward expected marginal revenue becomes

$$\text{Em} = \bar{p} \{\frac{1}{2} (1 - \Pi)(1 - b + 1 + b) + \Pi (1 - \frac{1}{\varepsilon})\}$$

$$= \bar{p} (1 - \Pi/\varepsilon)$$

instead of $\bar{p} (1 - \frac{1}{\varepsilon})$. The monopolist will thus store if his current harvest, h, satisfies

$$1 - \varepsilon b < h < 1 + \varepsilon b$$

(i.e. there is no buffer stock intervention), and

$$1 + \frac{1}{\varepsilon} - \frac{2h}{\varepsilon} + k/\bar{p} \quad < \quad \frac{\beta \text{Em}}{\bar{p}} = \beta (1 - \Pi/\varepsilon)$$

or

$$1 + \varepsilon b > h > \frac{1}{2} \{1 + \beta \Pi + \varepsilon (r + c)\} \qquad (30)$$

which is a very weak condition indeed (Note that Π depends on b). Thus, if $\beta = 0.95$, c = 3 percent, h is roughly normal with CV of 24 percent, $\varepsilon = 3$, then a competitive Bandwidth Rule would require $b \leqq 8$ percent, but if b is set at 8 percent, (so that $\Pi = 0.67$), then the monopolist will store on his own behalf whenever $0.94 < h < 1.24$, i.e. 35 percent of the time.

It is clear that the same problem will arise even when the consumer demand schedule is non-linear. Thus, if an international buffer agency is set up and instructed to maintain prices within a bandwidth, then any monopolist will probably (depending on the various parameters) be able to speculate against the agency, buying whenever the price falls near, but not quite as far as, the floor price in order to sell the following period (or periods).

The Competitive Rule

Suppose, however, that the buffer agency is instructed to follow the competitive storage rule (i.e., the rule relating storage to supply which would prevail in a competitive market). Such a rule would have two obvious rationales - it would break even

if there were constant storage costs (or make a profit
with increasing marginal costs), and it would be
efficient given the already mentioned assumptions of
risk neutrality etc. It would have the additional
advantage when confronting a monopolist subject to
stochastic supply shocks of being non-manipulable.
There are two possible cases, depending whether the
monopolist would undertake more or less stabilization
than a competitive producer. If he would undertake
more (e.g. the demand schedule is linear) then the
buffer agency will never intervene, since the remaining
price variability would be inadequate to justify
competitive intervention. If he would undertake less,
(e.g. if he faced a constant elasticity demand schedule)
then the agency would have a comparative advantage in
storage, as discussed above in section 4, and would
undertake all storage, with the monopolist performing
no storage.

There remains one curiosum, which is a consequence
of the kink induced in the effective demand schedule at
the point at which the agency first adds to storage
(x_0 - S in Figure 3). Newbery (1978) demonstrated
that it would in general pay a monopolist facing a
kinked demand schedule (one which becomes more elastic
just below the kink) to introduce randomness into an
otherwise non-random market. The reason is that the
monopolist can exploit the discontinuity in the marginal
revenue facing him. (Contrast the kinked demand theory
of oligopoly, in which the demand schedule becomes
less elastic below the kink, which provides an
explanation for price stability.)

COUNTERVAILING INTERNATIONAL PRICE STABILIZATION

If some commodity markets are recognized to be
monopolised or cartellised, then instead of following
a passive role, more suited to a competitive market,
the international stabilization agency might be set up
to try and mitigate the monopolistic distortions. At
this stage it should be pointed out that the political
impetus behind the call for an Integrated Programme
for Commodities (by UNCTAD in Nairobi, May 1976) was
more interested in countervailing the imperfections on
the buying side by creating producer cartels, rather
than defending the consumers against monopolistic
producers, and it would be interesting to extend the
analysis to consider monopsony importers facing
competitive producers, but that will have to await
another occasion. Some of the insights obtained by
studying the more tractable monopoly producer case may,
however, be useful.

This case has already been the subject of a most
interesting study by Nichols and Zeckhauser (1977).

They considered the case of a large consuming nation
(such as the U.S.), containing many competitive final
consumers, facing a cartel (obviously OPEC provided the
motive for addressing the question, but they drew
attention to cartels in bananas, bauxite, coffee, copper,
iron ore, mercury, phosphate, and tin). They
demonstrated that under certain circumstances it would
pay the consuming nation to build up a strategic
stockpile, whose presence would suppress prices in
future periods, even when the supply conditions of the
producing cartel were non-random and stationary, so
that on competitive price arbitrage grounds there would
be no case for stockpiling.

The reference model had the following features.
The cartel maximizes the present discounted revenues
net of production costs by setting the supply price
each period. The consuming nation maximizes the
present value of net consumers' surplus (consumers'
surplus less storage and interest charges on the
stockpile), by choosing its stockpile level each period.
Consumers decide on current consumption given the
current price. In the simplest model the world lasts
two periods, in the first of which the consumer
government buys stocks for resale next period. The
game is one of full information in which the supplier
plays first (announcing a price) and the consumer
government plays second (choosing a stockpile level).
The game is solved recursively, and can be solved
numerically for any finite number of periods. With
zero production and storage costs the effect of the
stockpile was to make future demand more elastic and
hence reduce the deadweight loss of the monopoly.
Consumers benefit from this, whilst with linear demand
and two periods, producers also gain because of the
higher first period demand and price. In short, both
parties gain from the countervailing market power
which reduces deadweight inefficiency losses, though as
the number of time periods increase, so the consumers
gain relatively to producers.

These results are quite striking, but need careful
interpretation. First, the benefits are measured by
comparing an interventionist stockpiling strategy by
the sole consuming country with no intervention at
all. The fact that the country is the sole consumer
means that the market structure is one of potential
bilateral monopoly, rather than pure monopoly, and it
might have been more logical to compare stockpiling
with other countervailing actions such as import tariffs.

Second, the cartel's supply strategies are rather
special, and correspond more closely to OPEC's
strategies than those of an agricultural commodity
cartel (though there are additional complicating factors
in the case of exhaustible resources which substantially

modify the conclusions, as the authors recognize). Only if short run supply is highly elastic will it be feasible for a cartel to announce a supply price and allow consumers then to dictate demand and hence supply. If this elasticity of supply is to be achieved by producer stockpiling then the same complicating intertemporal issues arise as with exhaustible resources, and the analysis must be substantially modified. It is clear that this assumption affects the relative market powers of the two parties substantially. If, for example, the producer always has the first say, and specified a supply price, then the importer can never usefully impose an import tariff, since this, by assumption, will not affect the supply price. Clearly, if the importer plays first, and announces a tariff, then the roles are reversed and the producer faces a different demand curve against which he optimises, and the resulting market equilibrium is quite different. This is dramatically illustrated in the following numerical example, which uses the same model as Nichols and Zeckhauser (1977, pp. 70-74).

Tariffs vs. Stockpiles in the Nichols-Zeckhauser Model

We employ the same notation and format as the two period model. Consumption, C, and price, p, are related by

$$C = K - \alpha p,$$

the discount factor is β, and there are no production or storage costs. With no consumer action, the price is set at the monopoly level $p^m = K/2\alpha$, and consumers enjoy a present value of consumer surplus $V^m = K^2(1 + \beta)/8\alpha$. If the sequence of events is (1) cartel selects p_1, (2) consumers select $C_1(p_1)$ and their government chooses stock, S, (3) cartel selects p_2 (4) consumers select $C_2(p_2)$ and import $C_2 - S$, then Nichols and Zeckhauser compute the optimum p_1, p_2 and S. With $\beta = 0.9524$ (rate of interst = 5 percent) they show that the optimal stockpiling strategy raises the net present value of consumer surplus by 7.7 percent and the cartel's revenue rises 6 percent.

If, in contrast, the sequence is (1) cartel selects p_1, (2) consumers choose $C_1(p_1)$, (3) the government sets an ad valorem tarrif τ for the second period (4) the cartel selects p_2, and (5) consumers demand $C_2\{p_2(1 + \tau)\}$, then it is easy to show that the optimum tariff is $\tau = 100$ percent, in which case the present value of consumers' surplus $K^2(1 + 2\beta)/8\alpha$, or an increase of 49 percent over the no intervention case. The cartel loses 24.4 percent of its present discounted revenue, though it is interesting to note that there is no change in deadweight loss, merely a redistribution from producers to consumers.

The main conclusion to draw from this fascinating study of the countervailing power of importer's stockpiling strategy is that the benefits of any such strategy depend very much on the initial and final market equilibria. With bilateral monopoly different solution concepts (i.e. different restrictions on allowable strategies, including their timing) have very different implications for the distribution of benefits and the magnitude of deadweight losses. Thus allowing one or other players to increase his range of allowable strategies to include stockpiling (with its attendant implications for the location and elasticity of supply and demand schedules within any period) can be expected to change the equilibrium. However, without knowing what constrains the original choice of strategies it is difficult to say who would benefit from stockpiling (i.e. who would undertake it), and also it is difficult to say what form optimal intervention would take.

NOTES

1. Why, for instance, are farmers protected at the expense of consumers only when they are a minority?

2. This assumption can be relaxed as shown in Newbery (1982).

REFERENCES

Bale, M.D., and Lutz, E. "Price Distortions in Agriculture and Their Effects: An International Comparison". American Journal of Agricultural Economics. 63 (1981): 8-22.

Boadway, R.W. "The Welfare Foundations of Cost-Benefit Analysis". Economic Journal. 84 (1974): 926-39.

FAO "Food Reserve Policies for World Food Security: A Consultant Study of Alternative Approaches". EJC:CSP/75/2, Rome, Food and Agriculture Organization. 1975.

Gray, R.W. "The Characteristic Bias on Some Thin Future Markets". Food Research Institute Studies. Nov. (1960).

Gustafson, R.L. Carryover Levels of Grains, U.S. Department of Agriculture. Technical Bulletin No. 1178. 1958.

Harberger, A.C. "Three Basic Postulates for Applied Welfare Economies: An Interpretive Essay". Journal of Economic Literature. 9 (1971): 785-97.

Hart, O.D. "On The Profitability of Speculation".
    Quarterly Journal of Economics, 91 (1977): 579-97.
Johnson, D. Gale. World Food Problems and Prospects,
    Washington, D.C., American Enterprise Institute
    for Public Policy Research. 1975.
Lutz, E. and Scandizzo, P.L. "Price Distortions in
    Developing Countries: A Bias Against Agriculture".
    European Review of Agricultural Economics. 7 (1980):
    5-27.
Newbery, D.M.G. "Stochastic Limit Pricing". Bell
    Journal of Economics. 9 (1978): 260-9.
Newbery, D.M.G. "Commodity Price Stabilization in
    Imperfect or Cartelized Markets". Mimeo, DRD,
    World Bank, Washington, D.C. 1982.
Newbery, D.M.G. and Stiglitz, J.E. The Theory of
    Commodity Price Stabilization. Oxford University
    Press. 1981.
Newbery, D.M.G. and Stiglitz, J.E. "The Choice of
    Techniques and the Optimality of Market Equilibrium
    with Rational Expectations". Journal of Political
    Economy. 90 (1982a): 223-246.
Newbery, D.M.G. and Stiglitz, J.E. "Optimal Commodity
    Stockpiling Rules". Oxford Economic Papers.
    (1982b): 403-427.
Nichols, A.L. and Zeckhauser, R.L. "Stockpiling
    Strategies and Cartel Prices". Bell Journal of
    Economics. 8 (1977): 66-96.
Salant, S.W. "The Vulnerability of Price Stabilization
    Programs to Speculative Attack", Mimeo. 1979.
Samuelson, P.A. "Stochastic Speculative Price".
    Proc. Nat. Acad. Sci. 68 (1971): 335-7.
Spence, A.M. "Entry, Capacity Investment, and
    Oligopolistic Pricing". Bell Journal of Economics
    8 (1977): 534-44.

12
Storage Under a Cartel

Leslie Young and Andrew Schmitz

INTRODUCTION

Most studies of the welfare effects of buffer
stock stabilization schemes have assumed competitive
behaviour by consumers and producers: Waugh (1944),
Oi (1961), Massell (1969), Samuelson (1972), Hueth and
Schmitz (1972), Turnovsky (1976), and Newbery and
Stiglitz (1979). However, production is cartelised in
many of the markets for which buffer stocks have been
mooted. This paper considers the welfare effects of
storage when production is controlled by a cartel which
behaves as a monopolist.

Our interest in the welfare effects of storage
under a cartel was stimulated by a puzzling feature of
the milk market in the United States. Because of the
existence of large producer cooperatives and marketing
orders, milk production is non-competitive (Ippolito
and Masson, 1978). Also, during a given year,
production fluctuates greatly, as does demand, with
the high demand occurring during the high cost, low
production periods (USDA, 1980). The technology exists
to cope with these fluctuations by storing milk as a
powder, the reconstituted milk being essentially in-
distinguishable from fresh milk (USDA, 1980). Why
have producers lobbied effectively to block the
introduction of reconstituted milk into the market? In
a competitive industry, this might be explained by Oi's
theorem (Oi, 1961) that total profits are reduced by
the price stabilization that storage could bring about.
However, the milk producers are organized as a cartel,
and we shall see that the use of storage would increase
the cartel's profits.

Leslie Young, Department of Finance, University of
Texas, Austin and Andrew Schmitz, Agriculture and
Resource Economics, University of California, Berkeley.

We can offer two explanations for the milk producers' lobbying behaviour in the case where the optimal policy of the cartel under storage requires that different prices be charged in different periods. If storage were permitted, then arbitrageurs outside the cartel could profit by buying and storing in the low price period and re-selling in the high price period. If storage costs are negligible then this would tend to equalize prices over periods. Thus the cartel has to compare total profits without storage with total profits under storage with uniform pricing.[1] We shall see that profits can be lower in the latter situation. In addition, we will show how the existence of price supports, if set at high enough levels, can make storage non-profitable for the cartel. In the milk industry case, price supports have been in existence since the early 1930's.

Our analysis can easily be extended to international trade where cartels have played an important role. In addition to oil cartels, the proposals for a cartel in food and feedgrains have also been presented (Schmitz et al. 1981).

In the second to fifth sections we consider the effects of storage and uniform pricing on the cartel's profits using what is essentially Massell's two-period model with linear demand and marginal cost curves which are subject to additive shifts between periods. We also consider how storage by the cartel - both with and without uniform pricing - affects consumer welfare and aggregate welfare, as measured by consumer's surplus. As expected, storage increases aggregate surplus and the imposition of uniform pricing increases aggregate surplus further.

We next consider the sensitivity of our conclusions to the assumptions made about demand. In the sixth section we show that if the demand curve is subject to multiplicative shifts between periods then, under storage, the cartel chooses to set the same price in both periods. Hence, the first explanation given above for the cartel's attitude to reconstituted milk does not apply and some alternative explanation must be sought - such as the existence of price supports. In the seventh section we show that if the inverse demand curve is subject to multiplicative shifts then the use of storage reduces total output. This underlines the possibility that storage may reduce aggregate welfare, even though it increases the profits of the cartel.

THE MODEL WITH ADDITIVE SHIFTS IN DEMAND

We assume that there are two periods t = 1,2. The inverse demand curve in period t is:

$$p = a_t - bx \qquad\qquad t = 1,2 \qquad\qquad (1)$$

where p is price, x is quantity demanded and a_t and b are positive constants. The cost curve in period t is:

$$c(y) = c_t y + dy^2 \qquad\qquad t = 1,2 \qquad\qquad (2)$$

where y is output and c_t and d are positive constants.

We assume that $a_2 > a_1$ and $c_2 > c_1$, i.e., in period 2 both the marginal cost curve and the demand curve shift upwards (e.g., the milk industry case). We have assumed zero fixed costs in both periods but this is not essential for our conclusions.

In Figure 1, $A_t B_t$ represents the demand curve in period t and $A_t R_t$ represents the marginal revenue curve. The marginal cost curve in period t is:

$$c'(y) = c_t + 2dy$$

This is represented by the line $C_t D_t$ in Figure 1. If there is no storage, then the cartel maximizes profits by equating marginal revenue and marginal cost in each period. Thus in period t, the output x_t^n is determined by the condition:

$$a_t - 2bx_t^n = c_t + 2dx_t^n$$

or:

$$x_t^n = (a_t - c_t)/2(b+d) \qquad\qquad (3)$$

In Figure 1 this is represented by the distance Ox_t^n. The price OP_t^n in period t corresponds to the point Q_t^n on the demand curve vertically above X_t^n.

Now suppose that a storage activity is available. It would then be feasible for the cartel to choose any combination of sales x_t and outputs y_t in the two periods such that:

$$x_1 + x_2 = y_1 + y_2 \qquad\qquad (4)$$

If storage is costless then total profits are maximized by choosing the x_t and y_t so that the marginal revenue in both periods and the marginal cost in both periods are all equal to one another,[2] i.e.,

$$a_1 - 2bx_1 = a_2 - 2bx_2 = c_1 + 2dy_1 = c_2 + 2dy_2 \qquad (5)$$

The simultaneous solutions of (4) and the three equations in (5) can be depicted as follows.

In Figure 1, the horizontal line EF intersects the marginal revenue curves $A_t R_t$ at points with horizontal co-ordinates $Ox_t^s = x_t^s$ and intersects the marginal cost curve $C_t D_t$ at points with horizontal co-ordinates $Oy_t^s = y_t^s$. The fact that these intercepts all lie on

288

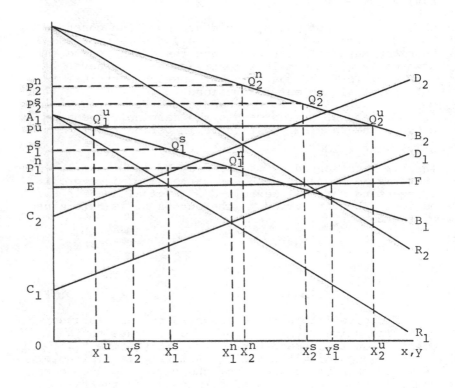

Figure 1 A Simple Model of Price Instability
          and Storage

the same horizontal line ensures that x_1^s, x_2^s, y_1^s, y_2^s
satisfy (5). The feasibility constraint (4) will be
satisfied if EF is drawn so that the mid-points of
x_1^s x_2^s and of y_1^s y_2^s coincide. The resulting prices OP_t^s
correspond to the points Q_t^s on the demand curve A_t B_t
vertically above X_t^s.
    Given the parallel shifts in the linear demand and
marginal cost schedules, the above construction implies
that the mid-points of Y_1^s Y_2^s and of X_1^n X_2^n coincide and
that the mid-points of P_1^s P_2^s and of P_1^n P_2^n coincide.

Thus the introduction of storage does not affect total output but shifts the production of Y_1^s X_2^n units from period 2 to period 1 and the consumption of X_1^s X_1^n units from period 1 to period 2. That is, with the introduction of storage, more of the good is both produced in the low-cost period and consumed in the high-demand period. Similarly, the introduction of storage does not affect the average price paid but reduces the fluctuations of prices.

An important alternative interpretation of the above model is that the curves C_t D_t are the industry supply curves in period t of competitive producers who sell their output to a marketing board. Suppose that the marketing board maximizes the sum of its own profits and those of the producers. This objective is in fact identical to that of the cartel described above since the costs of the marketing board net out the revenue received by the competitive producers. With this interpretation are conclusions that, under storage, the cartel should equate marginal costs between periods implies that the marketing board would set the same price to producers in both periods.[4] This could easily be accomplished by the use of quotas and pooled pricing--practises often used by marketing boards.

WELFARE EFFECTS OF STORAGE

Theorem 1: The introduction of storage increases the cartel's profits.
    Proof: This can be proven in terms of areas in Figure 1 but in fact follows from a "revealed preference" argument. The optimal choices of output and sales in the two periods when there is no storage will not equalize marginal revenue between periods nor marginal costs between periods (unless $a_1 = a_2$ and $c_1 = c_2$). Hence they represent feasible choices when storage is introduced but differ from the optimal policies under storage. Hence they yield lower total profits than the optimal policies under storage.

    For a cartel, the above argument is valid whatever the form of the demand and supply functions. By contrast, in the purely competitive case, prices are determined by industry-wide choices rather than by the choices of an individual firm. The above "revealed preference" argument would not be valid in this situation because, when a competitive firm is maximizing profits under storage, its individual choices could not bring about the prices faced by the industry in the absence of storage. In fact, for the competitive case, Massell (1969) showed [5] that the use of storage to stabilize prices will reduce profits if:

$$(a_2 - a_1)/(c_2 - c_1) > \sqrt{1 + b/d} \tag{6}$$

Theorem 2: If

$$(c_2 - c_1)/(a_2 - a_1) > 2 + d/b \tag{7}$$

then the use of storage by the cartel reduces consumer's surplus. If the inequality in (7) is reversed then consumer's surplus is increased.

Proof: If period 1, storage decreases consumer's surplus by the area P_1^n P_1^s Q_1^s Q_1^n (Figure 1). In period 2, storage increases consumer's surplus by the area P_2^n P_2^s Q_2^s Q_2^n. The construction of X_t^n, X_t^s, Y_t^s described in Section 2 ensures that these two trapezia have the same height. Moreover, Q_1^n Q_1^s and Q_2^n Q_2^s both have slope -b. Hence storage decreases consumer's surplus if and only if P_1^n Q_1^n > P_2^s Q_2^s or

$$x_1^n > x_2^s \tag{8}$$

To calculate x_2^s we use the equations in (5) to eliminate x_1, y_1 and y_2 from (4) and then solve for x_2^s. This yields:

$$x_s^2 = [2a_2 - c_1 - c_2 + d(a_2 - a_1)/b]/4(b + d) \tag{9}$$

By (8), (3) and (9), storage decreses consumer's surplus if and only if:

$$2a_1 - 2c_1 > 2a_2 - c_1 - c_2 + d(a_2 - a_1)/b$$

or

$$c_2 - c_1 > (a_2 - a_1)(2 + d/b)$$

If there are no shifts in demand, then P_1^n and P_2^n coincide and total consumer's surplus is decreased by storage. In this case, the above geometric argument resembles Waugh's argument (Waugh, 1944) that since consumer's surplus is convex in price, total surplus will decrease if the price distribution becomes more concentrated about its mean. When the demand curve shifts, storage confers a benefit on the representative consumer because the cartel then sells more in period 2 - when the consumer values the good more highly. Theorem 2 shows that the consumer will nevertheless be harmed if these demand shifts are small compared to the shifts in marginal cost. In the competitive case, Massell obtained a similar conclusion: the consumer is harmed by storage provided that:

$$(c_2 - c_1)/(a_2 - a_1) > \sqrt{1 + 4d/b} \tag{10}$$

This condition is less stringent than (7) since

$2 + d/b > \sqrt{1 + 4d/b}$.

Notice that if (7) were violated then storage benefits both consumers and producers. We would expect that even if consumers were harmed, they could be compensated by producers since storage leads the cartel to shift sales towards period 2 (where the good is valued more highly) and to shift production towards period 1 (where production costs are lower). The following argument makes this precise.

Theorem 3: The introduction of storage by the cartel increases aggregate surplus.

Proof: In period 1, storage decreases consumer's surplus plus sales revenue by the area Q_1^n Q_1^s X_1^s X_1^n (Figure 1). In period 2 storage increases consumer's surplus plus sales revenue by the area Q_2^n Q_2^s X_2^s X_2^n. The latter area is larger so storage increases consumer's surplus plus sales revenue.

In period 1, storage increases production costs by the area below C_1 D_1 between Y_1^s and X_1^n. In period 2, storage decreases production costs by the area below C_2 D_2 between Y_2^s and X_2^n. The latter area is larger so storage decreases total production costs. It follows that storage increases aggregate surplus - which equals consumer's surplus plus sales revenue less production costs.

We conclude this section by considering how our welfare conclusions would be modified if producers or consumers have non-constant marginal utility of income so that surplus measures fail to reflect changes in their welfare. From Figure 1 it can be seen that under storage the cartel produces more and sells less in period 1 than it would without storage. The reverse is true in period 2. Hence storage increases the variability of profits over the two periods. Despite this, it is unlikely that risk aversion explains the hostile attitude of the milk cartel toward storage since the fluctuations take place within a year and cartel members presumably have easy recourse to borrowing over this period.

For consumers with non-constant marginal utility of income, it can be shown that if the demand curve is fixed then storage by the cartel decreases total utility over the two periods if the coefficient of relative risk aversion ρ is less than the income elasticity of demand η, but can decrease expected utility if the reverse is true.[6] Further analysis of the case where the demand curve shifts between periods requires an explicit model of the source of the shift. Likewise, analysis of the effect of storage on aggregate welfare with non-constant marginal utility of income requires an explicit social welfare function.

STORAGE WITH UNIFORM PRICING

To analyze the optimal policy under storage with uniform pricing we construct the marginal cost curve for aggregate output when storage is used to allocate production efficiently between the two periods. In Figure 2 the marginal cost curve in period t is represented by the line $C_t D_t$ with vertical intercept C_t and slope 2d. Let W be the point on $C_1 D_1$ on the same horizontal level as C_2 and let $w \equiv C_2W$. If aggregate output $Y \leq w$ then total costs are minimized by producing only in period 2. Hence C_1W is the aggregate marginal cost curve over this range. For $y > w$, marginal increases of aggregate output are always allocated equally between the two periods (since total costs are minimized by equalizing marginal costs between periods). It follows that aggregate marginal costs increase with output at half the rate obtained for the marginal cost curves in either period. Hence, for $y > w$, the aggregate marginal cost curve has slope d as represented by WV in Figure 2.

WV intersects the vertical axis at the midpoint $\overline{C}$ of $C_1 C_2$ so its equation is:

$$c' = (c_1 + c_2)/2 - dy \qquad (11)$$

Therefore, the total costs of producing $y > w$ equals the area under $\overline{C}V$ to the left of y, less the area $C_1\overline{C}W$. The former area equals $(c_1 + c_2)y/2 - dy^2/2$ while the latter (viewed horizontally) is a triangle with base $(c_2 - c_1)/2d$ and height $(c_2 - c_1)/2d$ so its area is $(c_2 - c_1)^2/8b$. Hence, the total cost of producing $y > w$ is:

$$(c_1 + c_2)y/2 - dy^2/2 - (c_2 - c_1)^2/8d \qquad (12)$$

We also construct the demand curve for aggregate output when consumers face the same price in both periods. In Figure 2, $A_t B_t$ is the demand curve in period t, and T is the point on $A_2 B_2$ on the same horizontal level as A_1. Then the aggregate demand curve is A_2TU, where TU has a slope b/2 which is half that of the single-period demand curves. TU intersects the vertical axis at the midpoint $\overline{A}$ of $A_1 A_2$ so its equation is:

$$p' = (a_1 + a_2)/2 - bx/2 \qquad (13)$$

This gives the inverse aggregate demand curve in the range $x > A_1 T$.

The profit maximum under storage with uniform pricing occurs at the intersection of the aggregate marginal cost curve C_1WV and the marginal revenue curve which corresponds to the aggregate demand curve

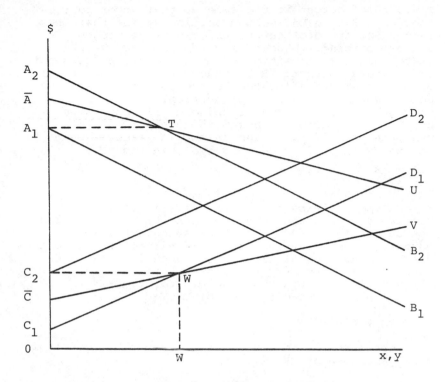

Figure 2   Storage with Uniform Pricing

A_2TU. Throughout the subsequent analysis, we assume that this intersection lies to the right of both W and T, i.e., under storage with uniform pricing, the cartel chooses to produce and sell positive amounts of output in both periods.

To depict the cartel's choices on Figure 1, we note that if aggregate sales and output are x then by (13) and (12) total profits are:

$$(a_1+a_2 - c_1-c_2)x/2 - (b+d)x^2/2 + (c_2-c_1)^2/8d \qquad (14)$$

The maximum of this expression occurs at:

$$x = x^u \equiv (a_1+a_2 - c_1-c_2)/2 \, (b+d) \qquad (15)$$

Comparing (15) and (3) we see that:

$$x^u = x_1^n + x_2^n$$

i.e., total output is the same as that under no storage. But in the second section we see that the latter equals total output under storage without uniform pricing. Thus if storage is already taking place then the imposition of uniform pricing does not affect total output. But, under storage, this total output is always allocated between periods so as to equate marginal costs of production. Hence under storage, the imposition of uniform pricing leaves unaltered the outputs chosen in each period. We conclude that if Y_t^u is the output in period t under storage with uniform pricing then, in Figure 1:

$$Y_t^u = OY_t^s$$

Sales in period t, x_t^u, are determined by the requirement that the sales in both periods be made at the same price p and that total sales equal total output. These requirements are depicted in Figure 1 by drawing a horizontal line $p^u Q_2^u$ whose intercepts Q_t^u with the demand curves $A_t B_t$ have horizontal coordinates OX_t^u such that the mid-point of $X_1^x X_2^u$ coincides with the mid-point of $Y_1^s Y_2^s$. Then:

$$x_t^u = OX_t^u, \; p^u = OP^u$$

This construction implies that p^u is the mid-point of $P_1^s P_2^s$ and also of $P_1^n P_2^n$.

WELFARE EFFECTS OF STORAGE UNDER UNIFORM PRICING

We consider how storage with uniform pricing affects the cartel's profits, consumer's surplus and aggregate surplus. In each case we compare these effects with what happens under storage without uniform pricing and under no storage.

Theorem 4: If the cartel is using storage then the imposition of uniform pricing decreases total profits.

Proof: This can be established using an obvious modification of the "revealed preference" argument of Theorem 1.

We next consider how total profits are affected by the move from no storage to storage with uniform pricing. Clearly, this reduces total costs since the production of $X_1^n Y_1^s$ units is shifted from the low cost to the high cost period. Against this, total revenue is reduced since uniform pricing prevents intertemporal price discrimination. The following result shows that the second effect dominates if the shift in demand between periods is large compared to the shift in marginal costs.

Theorem 5: If

$$(a_2 - a_1)/(c_2 - c_1) > \sqrt{1 + 1 + b/d} \tag{16}$$

then total profits are lower under storage with uniform prices than without storage. If the inequality (16) is reversed then the reverse conclusion holds.

Proof: Substituting from (15) into (14) we see that maximum profits under storage with uniform pricing is:

$$\pi u = \frac{(a_1 + a_2 - c_1 - c_2)^2}{8 \ (b+d)} + \frac{(c_2 - c_1)^2}{8d} \tag{17}$$

If there is no storage and x_t units are produced and sold in period t then profits in period t are:

$$(a_t - c_t)x_t - (b+d)x_t^2$$

As shown in (3), the maximum occurs at:

$$x_t^n = (a_t - c_t)/2(b+d)$$

and the maximum value of total profits is:

$$\pi^n = \frac{(a_1 - c_1)^2}{4(b+d)} + \frac{(a_2 - c_2)^2}{4(b+d)} \tag{18}$$

By (17) and (18), the introduction of storage with uniform pricing increases profits by:

$$\Delta\pi = [(c_2 - c_1)^2 \ (b+d)/d + (a_2 - c_2 + a_1 - c_1)^2 - 2(a_2 - c_2)^2$$

$$-2(a_1 - c_1)^2]/8(b+d)$$

i.e.:

$$\Delta\pi = [(c_2 - c_1)^2 \ (1 + b/d) - (a_2 - c_2 - a_1 + c_1)^2]/8 \ (b+d)$$

$\Delta\pi$ is negative if and only if:

$$\left[\frac{a_2 - a_1}{c_2 - c_1} - 1 \right]^2 = 1 + \frac{b}{d}$$

or

$$(a_2 - a_1)/(c_2 - c_1) > 1 + \sqrt{1 + b/d}$$

Thus if (16) holds then the cartel would attempt to block storage if it perceived that this would be accompanied by uniform pricing because of outside arbitrage or government regulation. However, the cartel would do even better by storing itself and

blocking storage by others and lobbying against uniform
pricing.

Theorem 6: If storage is taking place, then the
imposition of uniform pricing on the cartel increases
consumer's surplus.

   Proof: In period 1, the imposition of uniform
pricing decreases consumer's surplus by $P_1^s \, P^u \, Q_1^u \, Q_1^s$.
In period 2 it increases consumer's surplus by $P_2^s \, P^u$
$Q_2^u \, Q_2^s$. Since P^u is the mid-point of $P_1^s \, P_2^s$, the latter
area is larger.

Theorem 7: If

$$(c_2-c_1)/(a_2-a_1) > 3 + 2d/b \tag{19}$$

then consumer's surplus is lower under storage with
uniform pricing than with no storage. If inequality
(19) is reversed then the reverse will be true.

   Proof: In period 1, consumer's surplus under
storage with uniform pricing is less than that under
no storage by $P_1^n \, P^u \, Q_1^u \, Q_1^n$. In period 2, consumer's
surplus under storage with uniform pricing exceeds
that under no storage by $P_2^n \, P^u \, Q_2^u \, Q_2^n$. Since P^u is the
mid-point of $P_1^n \, P_2^n$, the latter area is smaller if and
only if:

$$P_2^n Q_2^n < P^u Q_1^u$$

or

$$x_2^n < x_1^u \tag{20}$$

Substituting from (15) into (13) we see that the price
set under storage with uniform pricing is:

$$P^u = (a_1+a_2)/2 - b(a_1+a_2-c_1-c_2)/4(b+d) \tag{21}$$

At this price the demand in period 1 is:

$$x_1^u = (a_1-p^u)/b$$

Substituting for p^u from (21) and using (3) we see that
$x_2^n < x_1^u$ if and only if:

$$\frac{a_2 - c_2}{2(b+d)} < \frac{a_1 - a_2}{2b} - \frac{c_1 + c_2}{4(b+d)} + \frac{a_1 + a_2}{4(b+d)}$$

or

$$\frac{a_2 - a_1}{(b+d)} + \frac{2(a_2 - a_1)}{b} < \frac{c_2 - c_1}{(b+d)}$$

or

$$3 + 2d/b < (c_2 - c_1)/(a_2 - a_1)$$

If storage is taking place then the adoption of uniform pricing means that the consumption of X_1^s X_1^u units is shifted from period 1 to period 2. Since the consumer values the food more highly in period 2 this increases consumer's surplus. Naturally, this means that the condition (19) for the consumer to be harmed by the move from non-storage to storage with unifirm pricing is more stringent than the condition (7) which applies when storage without uniform pricing is adopted.

Theorem 8: If storage is taking place, then the imposition of uniform pricing on the cartel increases aggregate surplus.

   Proof: In period 1, the imposition of uniform pricing decreases consumer's surplus plus sales revenue X_1^u X_1^s Q_1^u Q_1^s. In period 2 it increases consumer's surplus plus sales revenue by X_2^s X_2^u Q_2^u Q_2^s. The latter area is larger. In section 4 we saw that production costs are the same under both regimes. Hence aggregate surplus is increased by the imposition of uniform pricing.

Theorem 9: Aggregate surplus is higher under storage with uniform pricing than under no storage.

   Proof: This follows immediately from Theorems 3 and 8.

   Thus if a cartel is engaging in storage but manages to block outside arbitrageurs and price discriminates over time then the government can improve aggregate welfare by enforcing a uniform price. Moreover, the cartel may lose from a move from no storage to storage with uniform pricing, but this loss is always outweighted by the gains to consumers from this move.

EFFECT OF STORAGE ON PROFITS UNDER MULTIPLICATIVE UNCERTAINTY IN DEMAND

   In the second, third and fourth sections we assumed additive shifts in demand, i.e, that at every price, the demand in period 1 is less than that in period 2 by a fixed amount. Turnovsky (1976 p. 133) has argued that the assumption of multiplicative shifts in demand is often preferable on both theoretical and empirical grounds. In this case, the demand in period t can be written:

$$x_t = \theta_t f(p) \tag{22}$$

where $f(p)$ is a function of price alone and θ_t is a parameter specific to period t. Thus demand in period 1 is a fixed fraction of demand in period 2 at all prices.[7]

Theorem 10: Suppose that the inter-period shifts in

demand have the multiplicative form (22) and that in both periods, marginal revenue increases with price. Then the profit-maximizing policy for the cartel using storage involves setting the same price in both periods.

Proof: At price p, the elasticity of demand in period t is

$$\epsilon(p, \theta_t) = -p\theta_t(df/dp)/\theta_t f(p) = -p(df/dp)/f(p)$$

Thus under multiplicative shifts, the demand elasticity depends only on the price p and not on the period t being considered. Since the marginal revenue in period t is

$$p(1-1/\epsilon(p, \theta_t))$$

this also depends only on the price and not on the period t being considered.

If profits are to be maximized with storage then the marginal revenue must be the same in both periods (or else profits could be increased by re-allocating sales between periods using storage). But our hypotheses imply that, in both periods, marginal revenue is the same monotonic function of price. Hence the same price is set in both periods.

With multiplicative shifts in demand, the cartel using storage would choose uniform pricing. Our "revealed preference" argument then shows that its profits are always higher under storage with uniform pricing than under no storage. Thus, to explain a cartel lobbying against storage we would have to advance some argument other than that given after Theorem 5.

In the case of the milk producers, the explanation may be the existence of price supports in period 1 (the low demand, low cost period). The argument is illustrated in Figure 3 (where the labelling is similar to Figure 1). Without storage, the cartel's profits in period 1 equal the area $P_1^n Q_1^n J C_1$. If the price is supported at the level OP^h then the cartel could obtain profits $C_1 K P^h$ in period 1 by producing an output ON (and incidentally a surplus MN). It would lose revenue $P_1^n P^h I Q_1^n$, due to the lower price but would gain an additional rent IJK. Clearly, the cartel could be better off producing output ON, even if the supported price OP^h were less than the monopoly price OP_1^n.

With a sufficiently high level of price supports in period 1, profits would be higher without storage than that under the otpimal policy with storage but with no price supports. The cartel would then seek to prevent storage if it feared that this would be accompanied by the removal of price supports. This is plausible: supports are beneficial to the producer only if they lead to a surplus and, once storage is admitted, it may

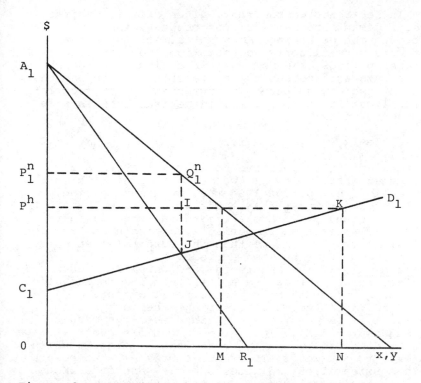

Figure 3   Instability and Multiplicative Demand
          Disturbances

become politically difficult to defend policies which
result in an excess supply taken over the whole year.
Thus we have an explanation for producer opposition to
storage for the case of multiplicative shifts in demand
(despite the implications of Theorem 10) and also for
the case of additive shifts in demand when condition
(7) is violated.

     If we interpret the model as that of a producer
marketing board (as explained at the end of Section 2)
then Theorem 10 implies that, under multiplicative
uncertainty in demand, the marketing board uses storage
to stabilize the prices faced by both consumers and
producers - but at different levels.

EFFECT OF STORAGE ON TOTAL OUTPUT UNDER MULTIPLICATIVE
UNCERTAINTY IN THE INVERSE DEMAND CURVE

In Section 3 we saw that, given additive shifts in
linear demand and marginal cost curves, the
introduction of storage does not affect total output
but merely re-allocates consumption and production
between periods. We now exhibit a situation where the
use of storage reduces aggregate output. Suppose that
the inverse demand curve is linear and is subject to
multiplicative shifts, i.e., in period t the inverse
demand curve is:

$$p = \phi_t \ (e - f x_t) \qquad t = 1, 2 \qquad (23)$$

where e and f are positive constants and ϕ_t is a
parameter specific to period t. In keeping with our
convention that period 1 is the low demand period, we
shall assume that $\phi_1 < \phi_2$. Then, in period 1, both the
inverse demand curve and the marginal revenue curve
will be flatter than in period 2, as shown in Figure 4.
For simplicity we assume that the marginal cost curve
CD does not shift between periods.

Theorem 11: If the inverse demand curve is linear and
subject to multiplicative shifts then the introduction
of storage reduces the cartel's total output.

Proof: If there is no storage then output in
period t is $OX_t^n = x_t^n$ where CD interests the marginal
revenue curve $A_t R_t$. The cartel's choices under storage
can be depicted by drawing a horizontal line EF which
intersects the marginal revenue curves $A_t R_t$ at points
with horizontal co-ordinates $OX_t^s = x_t^s$ and intersects
the marginal cost curve CD at the point with
horizontal co-ordinate $OY^s = y^s$. The fact that these
intercepts all lie on the same horizontal line ensures
that output and sales are such that the marginal
revenue in both periods and the marginal cost are all
equal to one another. Total output will equal total
sales provided that EF is drawn so that Y^s is the mid-
point of $X_1^s X_2^s$.

Since the marginal revenue curve in period 1, $A_1 R_1$,
is flatter than that in period 2, $A_2 R_2$, we conclude
that Y^s lies to the left of the mid-point of $X_1^n X_2^n$,
i.e.,

$$x_1^n + x_2^2 = OX_1^n + OX_2^n > 2(OY^s) = OX_1^s + OX_2^s = x_1^s + x_2^s$$

Under storage, marginal revenue and marginal cost will
be equated across periods. Lower total output will
result if marginal revenue decreases more quickly with
output in the high price period. This is because the
increase in output in the high price period will be
outweighed by the reduction in output in the low

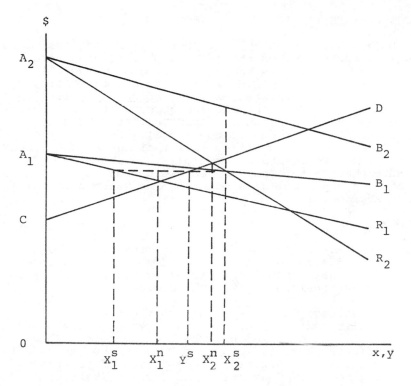

Figure 4 Storage with Multiplicative
          Uncertainty in Demand

price period which is necessary if the above marginal
condition is to be satisfied. It remains true that
storage tends to increase aggregate surplus by shifting
consumption towards the period where it is valued more
highly and by shifting production towards the period
where marginal costs are lower. Against this, however,
must be set the reduction in aggregate surplus
resulting from the reduction in total output.

Clearly, it is possible to draw demand and marginal
cost curves such that storage results in an overall
reduction of aggregate surplus. This is an example of
the General Theorem of Second Best. Storage removes
one "distortion" (the discrepancy between marginal
production costs in the two periods) but can so
exacerbate the effects of the other distortion (the

cartel's exercise of monopoly power) that aggregate welfare is reduced.

CONCLUSIONS

The analysis of the second to fifth sections yields sharp conclusions on the effects of storage on the welfare of producers and consumers and on overall welfare when production is cartelized. In particular, it suggests an explanation for producer hostility to storage and indicates that the government should permit storage and enforce uniform pricing, even when it accepts the existence of the cartel. However, the sixth and seventh sections show that these conclusions are sensitive to the assumptions made about demand and emphasize that policy recommendations need to be based on empirical work to determine which assumptions are appropriate.

NOTES

1. This would also be the case if the cartel believed that if storage were admitted when the government would enforce uniform prices by regulation - an impractical policy without storage.

2. If storage involves a unit cost g then, at a profit maximum, the marginal revenue (marginal cost) would differ by g between periods. This affects the details of our Theorems but not their qualitative character. The "revealed preference" arguments of Theorems 1 and 4 remain valid.

3. "Average" here means a simple average between periods, rather than a quantity-weighted average.

4. This result was noted earlier by Bieri and Schmitz (1974) but in a much simpler framework.

5. This can be seen if we interpret our marginal cost curve as his supply curve and recognize that $(a_2-a_1)/(c_2-c_1)$ is the ratio of the standard deviations of the heights of the demand and marginal cost curves.

6. This follows from Theorem 2 and Young and Anderson's results that utility is a concave (convex) function of consumer's surplus if $\rho > (<) \eta$ (Young and Anderson, 1981).

7. This occurs, for example, with additive shifts in the familiar loglinear model of demand.

REFERENCES

Bieri, J. and Andrew Schmitz   "Market Intermediaries and
    Price Instability: Some Welfare Implications",
    American Journal of Agricultural Economics,
    56, No. 2 (1974): 280-285.

Hueth, D. and Andrew Schmitz   "International Trade in
    Intermediate and Final Goods: Some Welfare
    Implications of De-stabilized Prices", Quarterly
    Journal of Economics, 86 (1972): 351-365.

Ippolito, R.A. and R.T. Masson   "The Social Cost of
    Government Regulations of Mills", Journal of Law
    and Economics, 21, (1978): 33-65.

Massel, B.F.   "Price Stabilization and Welfare",
    Quarterly Journal of Economics, 83, (1969):
    285-297.

Newbery, D.M.G. and J.E. Stiglitz   "The Theory of
    Commodity Price Stabilization Rules: Welfare
    Effects and Supply Responses", Economic Journal,
    89, (1979): 799-817.

Oi, W.Y.   "The Desirability of Price Instability Under
    Perfect Competition", Econometrica, 29, (1961):
    58-64.

Samuelson, P.A.   "The Consumer Does Benefit from
    Feasible Price Stability", Quarterly Journal of
    Economics, 86, (1972): 476-498.

Schmitz, A., A. McCalla, D. Mitchell and C. Carter
    Grain Export Cartels, Cambridge, Massachusetts,
    Ballinger Publication Co., 1979.

Turnovsky, S.J.   "The Distribution of Welfare Gains
    from Price Stabilization: The Case of
    Multiplicative Disturbances", International
    Economic Review, 17, (1976): 133-148.

U.S. Department of Agriculture, Agricultural Marketing
    Service "Handling of Milk in Federal Marketing
    Areas", Federal Register, 45:75955-75994, November
    17, 1980.

Waugh, Frederick V.   "Does the Consumer Benefit from
    Price Instability?" Quarterly Journal of
    Economics, 58, (1974): 602-614.

Young, Leslie and J.E. Anderson   "The Measurement of
    Welfare Benefits Under Price Uncertainty",
    University of Canterbury, 1981.